Synthesis Lectures on Mechanical Engineering

This series publishes short books in mechanical engineering (ME), the engineering branch that combines engineering, physics and mathematics principles with materials science to design, analyze, manufacture, and maintain mechanical systems. It involves the production and usage of heat and mechanical power for the design, production and operation of machines and tools. This series publishes within all areas of ME and follows the ASME technical division categories.

José I. Huertas
Editor

Fundamentals of Driving Patterns and Driving Cycles

 Springer

Editor
José I. Huertas
Tecnologico de Monterrey
Monterrey, Mexico

ISSN 2573-3168   ISSN 2573-3176 (electronic)
Synthesis Lectures on Mechanical Engineering
ISBN 978-3-031-76862-0   ISBN 978-3-031-76863-7 (eBook)
https://doi.org/10.1007/978-3-031-76863-7

This Springer imprint is published by the registered company Springer Nature Switzerland AG
The registered company address is: Gewerbestrasse 11, 6330 Cham, Switzerland

If disposing of this product, please recycle the paper.

Preface

The primary objective of this textbook is to present in an organized manner the knowledge accumulated worldwide regarding driving patterns and driving cycles (DCs). These two concepts are fundamental for improving the energy efficiency of vehicles, decarbonizing the transport sector, and promoting smart mobility in urban centers.

Through a systematic review of the literature published in the last 15 years, this document establishes a common language, offers a comprehensive review of the methodologies to construct DCs, identifies research opportunities, and presents recommendations on the best practices in the DC construction process that considers new trends and state of the art in technology.

This textbook results from an initiative of the Latin American Network for Research in Energy and Vehicles (RELIEVE, https://redrelieve.com/). RELIEVE is a non-profit organization that seeks to generate and strengthen capacities in its LATAM nodes, in a way that they could serve as technical support to government entities and private companies in topics related to energy, environment (air), vehicles, transport, and smart mobility. To achieve this objective, RELIEVE develops joint research projects, disseminates scientific knowledge, and trains human talent through specialized training and academic exchange among its participants.

More than 70 experts from 19 universities and four companies from seven Latin American countries actively participate in RELIEVE. Dr. Huertas from Tecnologico de Monterrey leads this organization.

The following table lists the main contributors to the present book.

List of authors

	Name	Last name	e-mail	Affiliation	Country
1	José I.	Huertas	jhuertas@tec.mx	Tecnológico de Monterrey	Mexico
2	Jenny	Díaz Ramírez	jdiaz@tec.mx	Tecnológico de Monterrey	Mexico
3	Michael Daniel	Giraldo Galindo	mgiral36@eafit.edu.co	Universidad EAFIT	Colombia
4	Luis Felipe	Quirama Londoño	luisfelipequirama@utp.edu.co	Universidad Tecnológica de Pereira	Colombia
5	Oscar Sebastián	Serrano-Guevara	oserrano@tec.mx	Tecnológico de Monterrey	Mexico
6	Daniel	Cordero-Moreno	dacorderom@uazuay.edu.ec	Universidad del Azuay	Ecuador
7	Luisa Fernanda	Chaparro Sierra	lchaparr@tec.mx	Tecnológico de Monterrey	México
8	Jessica Gisella	Maradey Lázaro	jmaradey@unab.edu.co	Universidad Autónoma de Bucaramanga	Colombia
9	Nicolas	Giraldo Peralta	nicolas.giraldo@uan.edu.co	Universidad Antonio Nariño	Colombia
10	Rogelio	Escamilla	A00838254@tec.mx	Tecnológico de Monterrey	México
11	Franco	Quezada	fnquezadab@gmail.com	Universidad de Concepción	Chile
12	Felipe	Vásquez	fevasquez@udec.cl	Universidad de Concepción	Chile
13	Jairo	Castillo Calderón	jdcastilloc@unl.edu.ec 894645@unizar.es	Universidad Nacional de Loja Universidad de Zaragoza	Ecuador España

(continued)

(continued)

	Name	Last name	e-mail	Affiliation	Country
14	Efren	Fernández	efernandez@uazuay. edu.ec	Universidad del Azuay	Ecuador
15	Danilo	Dávalos	danilodava_7@ hotmail.com	Universidad del Azuay	Ecuador
16	Freddy	Vasquez	fnvazquez@uazuay. edu.ec	Universidad del Azuay	Ecuador
17	Juan Danilo	Molina	Jmolina54@unab.edu. co	Universidad Autónoma de Bucaramanga	Colombia
18	Juan Carlos	Castillo	jccastillh@eafit.edu. co	Universidad EAFIT	Colombia
19	Victor	Romero	romerocanov@ cardiff.ac.uk	Cardiff University	UK
20	Juan Pablo	Jiménez	A01281906@tec.mx	Tecnológico de Monterrey	México

Monterrey, Mexico   José I. Huertas

Acknowledgments The authors thank the Latin American Network for Research in Energy and Vehicles (RELIEVE) (Ref. 720RT0014), whose operation is partially financed by the Ibero-American Program of Science and Technology for the development (CYTED). The authors also thank Dr. Omar Lopez, from Universidad de Los Andes, Colombia, for promoting the initiative of collecting the knowledge accumulated by RELIEVE on the topics of driving patterns and driving cycles.

Contents

Introduction

1

Jenny Díaz Ramírez⬤ and José I. Huertas⬤

Abstract

A driving cycle is a time series of speeds that represent a driving pattern, which in turn describes the typical driving behavior observed in a specific region. The construction of driving cycles (DCs) has been a significant research focus due to their role in evaluating vehicle energy demands and pollutant emissions, and more recently, in optimizing energy management systems for electric vehicles (EVs). This chapter introduces the essence and purpose of the book 'Fundamentals of Driving Patterns and Driving Cycles' and provides an overview of its content.

1.1 Introduction to Fundamentals on Driving Patterns and Driving Cycles

A driving pattern "describes how people drive in a given region", while a driving cycle (DC) is "a time series of speeds representing that driving pattern" [1–3]. DCs are constructed mainly (i) to evaluate the vehicle's energy consumption and emissions, (ii) as a tool to design vehicle components (e.g., powertrain), and (iii) evaluate energy management and logistics strategies in several applications involving motorized vehicles. The DC construction process has received much attention from the scientific community in the last

J. Díaz Ramírez (✉) · J. I. Huertas (✉)
Sustainable Energy Research Group, Tecnologico de Monterrey, Monterrey 64849, México
e-mail: jdiaz@tec.mx

J. I. Huertas
e-mail: jhuertas@tec.mx

© The Author(s), under exclusive license to Springer Nature Switzerland AG 2025   1
J. I. Huertas (ed.), *Fundamentals of Driving Patterns and Driving Cycles*, Synthesis
Lectures on Mechanical Engineering, https://doi.org/10.1007/978-3-031-76863-7_1

decades, looking for DCs that adequately represent the local driving pattern of the region or activity under study.

Through a systematic review of the literature published over the last fifteen years, combined with the expertise of the authors contributing to each chapter, this document establishes a common language, offers a comprehensive review of the methodologies to construct DCs, identifies research opportunities, and presents recommendations on the best practices in the DC construction process that considers new trends and state of the art in technology.

The development of DCs has been a research topic for the last 30 years. It is crucial for governmental environmental authorities and applied research centers focused on the energy usage and environmental footprint of vehicles [2]. In the contexts of smart cities, climate change, emerging vehicle technologies, and information technology, the interest in DCs worldwide has steadily grown. This interest has been fostered by the need to accurately identify DCs that allow the evaluation of the energy demands and pollutant releases from vehicles, as well as the optimization of energy management systems for electric vehicles (EVs) and powered hybrid EVs.

Despite its relevance, there is no standard method for constructing DCs and evaluating their degree of representativeness. In addition, there is a continuing demand for new DCs that meet specific requirements and for methodologies that easily allow the construction of personalized DCs from low-cost experimental data [4]. Progress in telematics, connected-vehicle systems, and big-data technologies has made it feasible and effortless to obtain data from vehicles to analyze driving characteristics, construct representative DCs, and develop efficient driving techniques and adaptive control strategies for EVs. Thus, a DC construction method with high accuracy and less experimental effort would be highly appreciated when dealing with big-data management [5].

In addition, the existing DCs cannot keep pace with the changing mobility dynamics: road infrastructure improvements, vehicle fleet composition change, the rising number of vehicles on the roads, and the focus on environment and resource protection. As a result, significant research has been conducted to develop new and representative DCs tailored to specific contexts [6, 7].

Although there is a progressive increase in quantity of publications on DC development, as shown later in Fig. 2.2a, there is a lack of an updated study that could offer, in a comprehensive way, a DC construction method. Most studies focus on specific steps of the process ignoring others that we consider fundamental to justifying the representativeness of a DC.

The primary objective of this work is to systematically present the knowledge accumulated in recent years concerning driving patterns and the DCs construction methods. The initial step involves establishing a common language related to driving patterns and DCs. Then, based on a systematic literature review, describe the steps required to construct a representative DC. We will also identify research trends and opportunities and, when appropriate, offer authoritative guidance on the best practices, considering contemporary technological advancements. Finally, seven practical cases of DC construction in various Latin American regions are presented, demonstrating the application of the construction process developed in this book.

References

1. Huertas, J., Giraldo, M., Quirama, L., Díaz, J.: Driving cycles based on fuel consumption. Energies (Basel) **11**, 3064 (2018). https://doi.org/10.3390/en11113064
2. Berzi, L., Delogu, M., Pierini, M.: Development of driving cycles for electric vehicles in the context of the city of Florence. Transp. Res. D Transp. Environ. **47**, 299–322 (2016). https://doi.org/10.1016/j.trd.2016.05.010
3. Tong, H.Y., Hung, W.T.: A framework for developing driving cycles with on-road driving data. Transp. Rev. **30**, 589–615 (2010). https://doi.org/10.1080/01441640903286134
4. Bender, F.A., Sawodny, O.: Development of a refuse truck driving cycle collective based on measurement data. Int. J. Environ. Waste Manag. **15**, 99 (2015). https://doi.org/10.1504/IJEWM.2015.068930
5. Shen, P., Zhao, Z., Li, J., Zhan, X.: Development of a typical driving cycle for an intra-city hybrid electric bus with a fixed route. Transp. Res. D Transp. Environ. **59**, 346–360 (2018). https://doi.org/10.1016/j.trd.2018.01.032
6. Yuhui, P., Yuan, Z., Huibao, Y.: Development of a representative driving cycle for urban buses based on the K-means cluster method. Cluster Comput. **22**, 6871–6880 (2019). https://doi.org/10.1007/s10586-017-1673-y
7. Schüller, M., Tewiele, S., Bruckmann, T., Schramm, D.: Evaluation of alternative drive systems based on driving patterns comparing Germany, China and Malaysia. Int. J. Autom. Mech. Eng. **14**, 3985–3997 (2017). https://doi.org/10.15282/ijame.14.1.2017.13.0323

Systematic Literature Review on Driving Cycles

Jenny Díaz Ramírez⬤, Jessica Gissella Maradey Lázaro⬤, and José I. Huertas⬤

Abstract

This chapter outlines the methodology used to conduct a systematic literature review on driving patterns and driving cycles. The objective is to ensure that the book includes the most significant contributions to this area of knowledge. We also identify the most frequently cited authors and active research networks in this field. Additionally, we provide a comprehensive summary of driving cycles constructed in various regions worldwide, highlighting the data collection techniques and primary construction methods employed.

2.1 Systematic Literature Review

The most recent comprenhensive literature review on DC construction methods was published over a decade ago [1], with additional reviews available in [2, 3]. To ensure the value of documenting each decision made during the review process and to make the results both useful and replicable [4], the systematic literature review methodology illustrated in Fig. 2.1 was adopted, based on the approaches detailed in [5–7].

J. Díaz Ramírez (✉) · J. I. Huertas
Sustainable Energy Research Group, Tecnologico de Monterrey, Monterrey 64849, México
e-mail: jdiaz@tec.mx

J. I. Huertas
e-mail: jhuertas@tec.mx

J. G. Maradey Lázaro
Universidad Autónoma de Bucaramanga, 680003 Bucaramanga, Colombia
e-mail: jmaradey@unab.edu.co

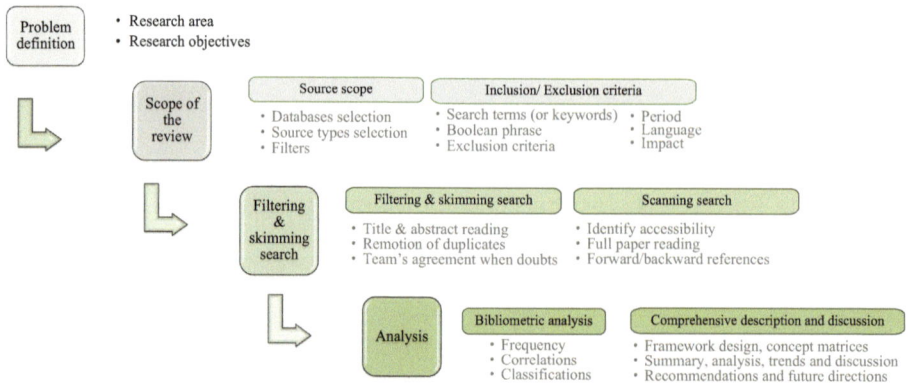

Fig. 2.1 Description of the methodology employed in this study for conducting the systematic literature review on driving patterns and DCs

The first stage of the systematic literature review is the "problem definition." In this case, the methodologies to describe driving patterns and construct driving patterns. Then, the scope of the literature review is defined. In this case, it encompassed the following considerations:

- The search was limited to the Scopus and Web of Science (WoS) databases.
- The Boolean phrase used was: ("Driving cycles" OR "driving cycle" OR "operation cycle" OR "driving pattern" OR "driving patterns") AND (develop* OR construct* OR build* OR synthesis OR framework OR select* OR generat* OR detect* OR comput* OR predict* OR optim* OR estimat* OR evaluat*) AND (typical OR representat* OR real-world OR city OR cities OR region* OR vehicle* OR bus* OR truck* OR motorcycle* OR electric* OR road* OR hybrid*). The first conjunction sentence was looked at in titles, and the last two were in Topics for WoS and ABS-TIT-KW for Scopus).
- A period between January 2005 and November 2023 was considered.
- Only published articles and proceedings in the English language were included.

Following the searching and filtering stage, from 867 results from Scopus and 321 from WoS, there were 859 different documents (set A in Fig. 2.2a). We removed duplicates of the same article in both databases (i.e., same authors with a related publication in more than one source. e.g., the first paper is in proceedings and the second in a journal. In these cases, for example, we kept the journal version).

Next, we conducted a second round of filtering by reviewing the abstracts and key-words, applying the exclusion criterion that the paper must address the construction or development of a DC rather than merely its use or evaluation. Any papers that were uncertain were reviewed by a peer, and their inclusion in the corpus was determined by consensus. This process resulted in the identification of 123 relevant documents (sets C1 and C2 in Fig. 2.2a).

2.2   A Glance of the Research on Driving Cycles

This section presents some highlights from the review process described in the previous section. The evolution of the last 14 years of research on DC development can be observed in Fig. 2.2a. The five papers from 2010 include four from the preceding years and are incorporated through the backward inclusion process. Figure 2.2a shows a consistently growing interest in the topic until 2019. The reduction in selected papers from 2020 onward may be attributed to two possible reasons: an unintended increase in rigor during the selection process, given the insights gained from the first searches, or as a sign that innovations in DC construction methods are reaching maturity.

A cluster analysis was performed to look for patterns. Results are presented in burble graphs as in Fig. 2.2b, which shows the country coauthor network. A strong coauthoring network inside China and a weaker relationship with other countries are shown. In addition, there is an interesting triad relationship between the US, Mexico, and Colombia. The rest of the coauthoring happens within each country. Over 90% of the corpus are Asian, mainly from China.

Figure 2.2c presents a coauthor co-citation network that shows a highly interconnected network among most countries, with China and the US being the more frequent nodes.

Figure 2.3a shows the frequency of papers per journal. Transportation Research Part D and Sustainable Cities and Society lead this topic with ~20% of the publications in the corpus, followed by 15 journals with 2–6 papers each, highlighting IEEE Access and Sustainability.

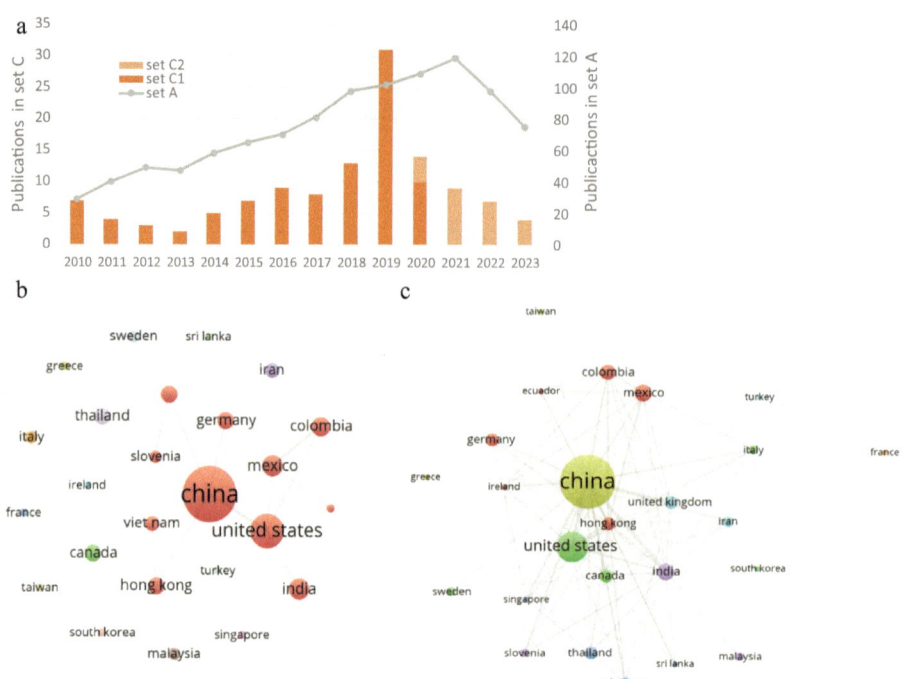

Fig. 2.2 Publications on DC construction method. **a** Frequency per year. **b** Coauthor country network. **c** Co-citation country network

Citation behavior (Fig. 2.3b) is consistent with the publication frequency observed in Fig. 2.3a, illustrating the prominence of Transportation Research Part D, Sustainable Cities and Society, IEEE Journals, and Applied Energy. The co-citation network in Fig. 2.3c shows three co-citation clusters, with journals such as Atmospheric Environment, Applied Energy, Energy, and IEEE Transactions in Vehicular Technology, being strongly connected.

In terms of authors, Fig. 2.3d shows a highly granulated participation of hundreds of authors up to 2020, with Hung and Tong being at the top, mainly because of their framework and review papers [1, 8], achieving 320 citations up to 2023 between the two. Table 2.1 shows the most cited group of authors, based on the Scopus database. Half correspond to one or two influencing papers published 8–17 years ago. At the same time, new groups of active researchers on the topic, such as Zhao' and Huertas', have emerged with 6 papers, primarily published in the last five years. Table 2.2 identifies the top 5 cited papers from the corpus.

Finally, a collection of papers dealing with the construction of DCs in specific regions is reported in Table 2.3. For future reference within this book, we underscore the following information: the geographic region for which the DC was built, and the types of vehicles

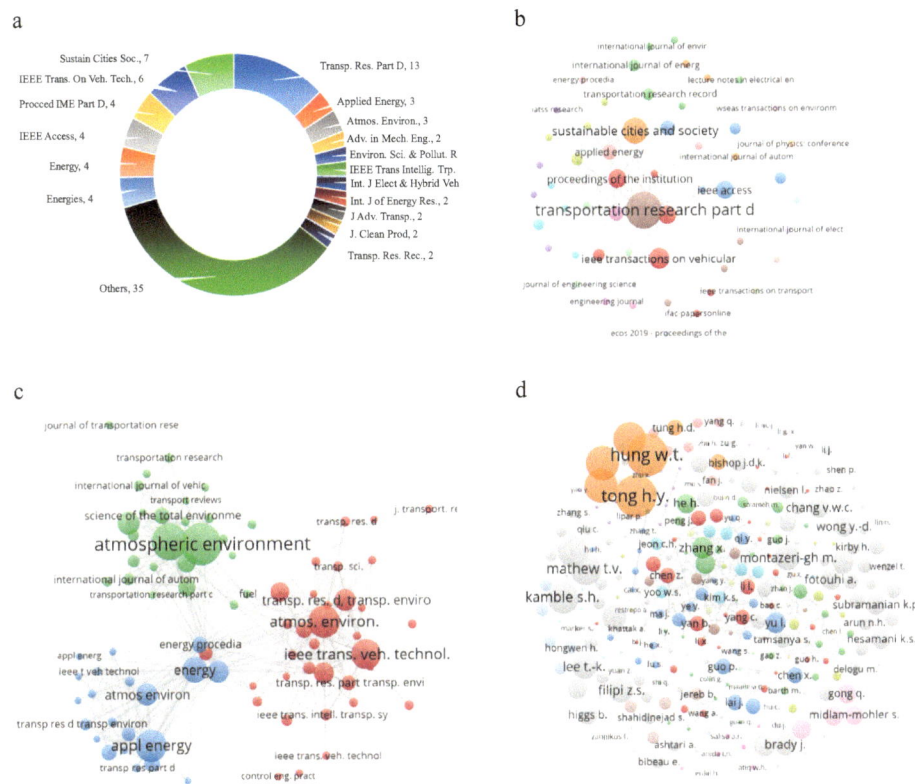

Fig. 2.3 Clustering analysis of published papers. **a** Publication frequency per source. **b** Citation network per source. **c** Co-citation network per source. **d** Author citation network

employed. These topics will be elaborated on in Chap. 3; the data collection techniques used in the process, which are discussed in Chap. 4; the construction methodology and DC duration, addressed in Chap. 5; and the quantity of characteristic parameters (CPs) considered in the construction process. CPs are discussed throughout the rest of the book.

2.3   Concluding Remarks

The methods for constructing driving cycles (DCs) have been a research focus for the past 30 years as they are crucial for evaluating vehicles' energy and environmental performance. A consistently growing interest in methods for constructing DC was observed until 2019. The reduction in the number of papers published from 2020 onward may suggest that innovations in DC construction methods are reaching maturity. Despite the

Table 2.1 The top ten most relevant groups of authors in the field

Authors	Documents	Cites[*]
Tong, H., Hung, W., et al.	6	437[**]
Zhao, X., Ye, Y., Ma, J. Wu, Y., et al.	6	332
Kamble, S.H., Mathew, T.V., & Sharma, G.K.	1	204
Lee, T.-K., Adonato, & Filipi	2	192
Brady, J., & O'Mahonyy, M.	1	175
Fotouhi & Montazeri	2	156
Huertas, J.I., Quirama, L.F., Giraldo, M., et al.	6	134
Ho, S.-H., Wong, Y.-D., Chang, V.W.C.	1	133
Nesamani, K.S., & Subramanian, K.P.	1	100
Lai, J.-X., Yu, L., Song, G.-H., Chen, X.-M., et al.	3	89
He, H., Sun, C., Zhang, X., et al.	3	73

[*] Cites of publications in the corpus by 24/01/24. [**] Excludes a paper from 1999 with 140 citations

Table 2.2 The top 5 most cited papers since 2005

Authors	Source	Year	Cites[*]
Hung, W.T., et al.	[8]	2007	231
Kamble, S.H., Mathew, T.V., & Sharma, G.K.	[9]	2009	204
Brady, J., & O'Mahonyy, M.	[10]	2016	175
Ho, S.-H., Wong, Y.-D., Chang, V.W.C.	[11]	2014	133
Nesamani, K.S., & Subramanian, K.P.	[12]	2011	100

[*] Cites of publications in the corpus by 24/01/24. [**] Excludes a paper from 1999 with 140 citations

relevance of the DCs and the apparent maturity of the field, no standard method exists for constructing DCs or assessing their degree of representativeness.

Ongoing research reveals a lack of consensus regarding their construction methods, definitions, and purposes. Efforts to refine or enhance DC-constructing methods continue, but no definitive standard for determining their suitability has emerged. Notably, there is a discernible shift towards developing DCs and construction methods tailored to specialized applications, such as Electric Vehicles (EVs), with a particular focus on incorporating machine learning techniques into the DCs construction process.

The next chapter will address these issues by outlining the core concepts on driving patterns and driving cycles.

Table 2.3 Summary of DCs developed for explicit regions

Region	Duration	Vehicle type	Data collection technique	DC construction methods	Number of CP's	Reference, Year
Fuzhou, China	~1200 s	B	OBD	MT	6	[4]
Fujian, China	1289 s	LV	GPS	TB/PCA/K-m	11	[13]
Shanghai, China	1200 s	HB	OBD	MT/PCA/K-m	8	[14]
Chennai, India	~1200s	LV, M	GPS/OBD	MT	11	[15]
Chennai, India	9.6 km	B	GPS/OBD	MT	14	[12]
Tianjin, China	1100 s	EV	OBD	LDA/MT/PCA	15	[16]
Kuala Terengganu	1254 s	LV	OBD	MT	9	[17]
Dublin, Ireland	1600 s	EV	GPS/ESB	MT/NN	10	[10]
Florence, Italy	3500 s	EV	OBD	MT/CA	19	[2]
Harbin, China	1600–1861 s	B	GPS	PCA/CA	6	[18]
Dalian, China	900 s	eB	GPS	PCA/CA	11	[19]
Khon Kaen, Thailand	1164 s	M	OBD	MT	9	[20]
Singapore	2344 s	LV	Chase-car	MT	20	[11]
Beijing, China	1084 s	BRT	GPS	MT	10	[21]
Tehran, Iran	1533 s	LV	GPS/ONDS	MT/K-m	2	[22]
Vietnam	2000 s	LV, M	OBD	MT	12	[23]
Pune, India	1501 s	LV	Chase-car/GPS	MT	6	[9]
Hong Kong	1600 s	LV	GPS/OBD	MT	14	[8]
Kaohsiung, Taiwan	4.3 km average	M	Chase-car	TB	11	[24]
Colombo, Sri Lanka	1200 s	LV	GPS	TB	2	[25]
Hamburg, Germany	~1200 s	B	OBD	MT	8	[26]
Bangalore, India	2088 s	LV	OBD	TB/MC/K-m	6	[27]
Hanoi, Vietnam	2061 s	B	OBD	Modal/MC	14	[28]
Ljubljana, Slovenia	1587 s	LV	OBD	MT	9	[29]
Winnipeg, Canada	1200 s	LV	OBD	Modal/MC	14	[30]
Mexico, Mexico	ND	B	GPS/OBD	MT	16	[31]
Kanchanaburi, Thailand	1120 s	B	OBD/GPS	MT	7	[32]
Malaysia	1138 s	LV	Chase car	MT	17	[33]
Celje, Slovenia	2453 s	LV	Chase car/GPS	ND	6	[34]
Hong Kong	~1600 s–~2800 s	B	OBD/GPS/ONDS	MT	13	[35]

(continued)

Table 2.3 (continued)

Region	Duration	Vehicle type	Data collection technique	DC construction methods	Number of CP's	Reference, Year
Fuzhou, China	Most: ~1750 s	LV	Chase car/ GIS	OBM/MC	9	[36]
Tehran, Iran	1800 s	LV	OBDII/GPS	MT	10	[37]
Xi'an, China	1200 s	EV	OBD/GPD/ ONDS	MC	9	[38]
Xiamen, China	1750 s	ERV	OBD/GPS	MT	7	[39]
Islamabad, Pakistan	1200 s	EV	OBD/GPS/ ArcGIS	MC	20	[40]
Hong Kong	1300 s, 1600 s	be	OBD/GPS/ ONDS	MT	13	[41]
Greater Cairo, Egypt	1600 s	LV	OBD/PEMS	MT	9	[42]
Zhangjiakou, China	1700 s	be	OBD/ ONDS	MC	9	[43]
Prishtina, Kosovo	1750 s	LV	OBD/CAN/ GPS	MT	8	[44]
Debrecen, Hungary	747 s	B	OBD	MT	3	[45]
Shenyang, China	1200–3000 s	EV	OBD/P-Box	MC/PCA/ CA/OBM	8	[46]
Xi'an, China	1200 s	EV	OBD/GPS/ V-Box	MT/PCA/ K-m/RF	14	[47]
Guandong, China	1581 s	FCV	OBD/ T-Box/GPS	MT/PCA/ CA	14	[48]
Zhengzhou, China	1200 s	eB	OBD/GPS/ ONDS	MC	16	[49]
Fuzhou, China	1200 s	LV	OBD/GPS	MT/PCA/ CA	14	[50]
Beijing, China	1800 s	B	OBD/ ONDS	MC/GA/ K-m	17	[51]
Guangzhou	1400–2000 s	LV	OBD/ GPS-PEMS	TB/PCA/ K-m	13	[52]
Istanbul, Turkey	RDC: ~500 s, Full: ~5500 s	BRT	OBD/CAN	MT	14	[53]
Edinburgh, Scotland	770 s urban, 656 s rural	M	Chase car/ OBD	TB	12	[54]
Montreal/ Slaberry-de-Valleyfield, Canada	1200 s	T	OBD	MT	10	[55]

EV: Hybrid/Electric Vehicle, LV: light or passenger vehicles, HV: hybrid vehicles, MC: Motorcycles, B: Buses, eB: electric buses, BRT: bus rapid transit, T: tracks or utility vehicles, FCV: fuel cell vehicle, ERV: electric ride-hailing vehicles, MT: Micro-trips, MC: Markov Chain/Monte Carlo, TB: Trip Segment Based, PCA: Principal Component Analysis, K-m: K-mean, CA: Other Cluster Analysis method, NN: Neural Networks, GA: Genetic Algorithm. ND: Not Determined, ESB: Electric Supply Board, ONDS: Other navigation or data systems, OBD: On-Board Device (implies On-Board technique), CAN bus: controller area network, GPS: Geographic Positioning System (or similar), OBM: other optimization-based method

References

1. Tong, H.Y., Hung, W.T.: A framework for developing driving cycles with on-road driving data. Transp. Rev. **30**, 589–615 (2010). https://doi.org/10.1080/01441640903286134
2. Berzi, L., Delogu, M., Pierini, M.: Development of driving cycles for electric vehicles in the context of the city of Florence. Transp. Res. D Transp. Environ. **47**, 299–322 (2016). https://doi.org/10.1016/j.trd.2016.05.010
3. Wang, Q., Huo, H., He, K., Yao, Z., Zhang, Q.: Characterization of vehicle driving patterns and development of driving cycles in Chinese cities. Transp. Res. D Transp. Environ. **13**, 289–297 (2008). https://doi.org/10.1016/j.trd.2008.03.003
4. Yuhui, P., Yuan, Z., Huibao, Y.: Development of a representative driving cycle for urban buses based on the K-means cluster method. Cluster Comput. **22**, 6871–6880 (2019). https://doi.org/10.1007/s10586-017-1673-y
5. Keathley-Herring, H., Van Aken, E., Gonzalez-Aleu, F., Deschamps, F., Letens, G., Orlandini, P.C.: Assessing the maturity of a research area: bibliometric review and proposed framework. Scientometrics **109**, 927–951 (2016). https://doi.org/10.1007/s11192-016-2096-x
6. Vom Brocke, J., Simons, A., Niehaves, B., Riemer, K., Plattfaut, R., Cleven, A.: Reconstructing the giant: on the importance of rigour in documenting the literature search process. In: Proceedings of the European Conference on Information Systems (2009)
7. Ruhlandt, R.W.S.: The governance of smart cities: a systematic literature review. Cities **81**, 1–23 (2018). https://doi.org/10.1016/j.cities.2018.02.014
8. Hung, W.T., Tong, H.Y., Lee, C.P., Ha, K., Pao, L.Y.: Development of a practical driving cycle construction methodology: a case study in Hong Kong. Transp. Res. D Transp. Environ. **12**, 115–128 (2007). https://doi.org/10.1016/j.trd.2007.01.002
9. Kamble, S.H., Mathew, T.V., Sharma, G.K.: Development of real-world driving cycle: case study of Pune, India. Transp. Res. D Transp. Environ. **14**, 132–140 (2009). https://doi.org/10.1016/j.trd.2008.11.008
10. Brady, J., O'Mahony, M.: Development of a driving cycle to evaluate the energy economy of electric vehicles in urban areas. Appl. Energy **177**, 165–178 (2016). https://doi.org/10.1016/j.apenergy.2016.05.094
11. Ho, S.-H., Wong, Y.-D., Chang, V.W.-C.: Developing Singapore driving cycle for passenger cars to estimate fuel consumption and vehicular emissions. Atmos. Environ. **97**, 353–362 (2014). https://doi.org/10.1016/j.atmosenv.2014.08.042
12. Nesamani, K.S., Subramanian, K.P.: Development of a driving cycle for intra-city buses in Chennai, India. Atmos. Environ. **45**, 5469–5476 (2011). https://doi.org/10.1016/j.atmosenv.2011.06.067
13. Zou, S., Zhu, X., Xu, L., Zhu, H.: Construction of vehicle driving cycle in Fuzhou. J. Phys. Conf. Ser. **1453**, 012146 (2020). https://doi.org/10.1088/1742-6596/1453/1/012146
14. Shen, P., Zhao, Z., Li, J., Zhan, X.: Development of a typical driving cycle for an intra-city hybrid electric bus with a fixed route. Transp. Res. D Transp. Environ. **59**, 346–360 (2018). https://doi.org/10.1016/j.trd.2018.01.032
15. Arun, N.H., Mahesh, S., Ramadurai, G., Shiva Nagendra, S.M.: Development of driving cycles for passenger cars and motorcycles in Chennai, India. Sustain. Cities Soc. **32**, 508–512 (2017). https://doi.org/10.1016/j.scs.2017.05.001
16. Jing, Z., Wang, G., Zhang, S., Qiu, C.: Building Tianjin driving cycle based on linear discriminant analysis. Transp. Res. D Transp. Environ. **53**, 78–87 (2017). https://doi.org/10.1016/j.trd.2017.04.005

17. Anida, I.N., Ismail, I.S., Norbakyah, J.S., Atiq, W.H., Salisa, A.R.: Characterisation and development of driving cycle for work route in Kuala Terengganu. Int. J. Autom. Mech. Eng. **14**, 4508–4517 (2017). https://doi.org/10.15282/ijame.14.3.2017.9.0356

18. Wu, X., Hu, C., Du, J.: Development of a driving cycle for city bus in Harbin of China. Int. J. Electr. Hybrid Veh. **7**, 104 (2015). https://doi.org/10.1504/IJEHV.2015.071063

19. Gao, X., Zhang, B., Xiong, X., Dong, M., Li, H.: Construction and analysis of the Dalian driving cycle. Int. J. Control Autom. **8**, 363–368 (2015). https://doi.org/10.14257/ijca.2015.8.6.35

20. Seedam, A., Satiennam, T., Radpukdee, T., Satiennam, W.: Development of an onboard system to measure the on-road driving pattern for developing motorcycle driving cycle in Khon Kaen city, Thailand. IATSS Res. **39**, 79–85 (2015). https://doi.org/10.1016/j.iatssr.2015.05.0030

21. Lai, J., Yu, L., Song, G., Guo, P., Chen, X.: Development of city-specific driving cycles for transit buses based on VSP distributions: case of Beijing. J. Transp. Eng. **139**, 749–757 (2013). https://doi.org/10.1061/(ASCE)TE.1943-5436.0000547

22. Fotouhi, A., Montazeri-Gh, M.: Tehran driving cycle development using the K-means clustering method. Sci. Iranica **20**, 286–293 (2013). https://doi.org/10.1016/j.scient.2013.04.001

23. Tong, H.Y., Tung, H.D., Hung, W.T., Nguyen, H.V.: Development of driving cycles for motorcycles and light-duty vehicles in Vietnam. Atmos. Environ. **45**, 5191–5199 (2011). https://doi.org/10.1016/j.atmosenv.2011.06.023

24. Tsai, J.-H., Chiang, H.-L., Hsu, Y.-C., Peng, B.-J., Hung, R.-F.: Development of a local real world driving cycle for motorcycles for emission factor measurements. Atmos. Environ. **39**, 6631–6641 (2005). https://doi.org/10.1016/j.atmosenv.2005.07.040

25. Galgamuwa, U., Perera, L., Bandara, S.: Development of a driving cycle for Colombo, Sri Lanka: an economical approach for developing countries. J. Adv. Transp. **50**, 1520–1530 (2016). https://doi.org/10.1002/atr.1414

26. Günther, R., Wenzel, T., Wegner, M., Rettig, R.: Big data driven dynamic driving cycle development for busses in urban public transportation. Transp. Res. D Transp. Environ. **51**, 276–289 (2017). https://doi.org/10.1016/j.trd.2017.01.009

27. Mayakuntla, S.K., Verma, A.: A novel methodology for construction of driving cycles for Indian cities. Transp. Res. D Transp. Environ. **65**, 725–735 (2018). https://doi.org/10.1016/j.trd.2018.10.013

28. Nguyen, Y.-L.T., Nghiem, T.-D., Le, A.-T., Bui, N.-D.: Development of the typical driving cycle for buses in Hanoi, Vietnam. J. Air Waste Manag. Assoc. **69**, 423–437 (2019). https://doi.org/10.1080/10962247.2018.1543736

29. Lipar, P., Strnad, I., Česnik, M., Maher, T.: Development of urban driving cycle with GPS data post processing. PROMET - Traffic Transp. **28**, 353–364 (2016). https://doi.org/10.7307/ptt.v28i4.1916

30. Ashtari, A., Bibeau, E., Shahidinejad, S.: Using large driving record samples and a stochastic approach for real-world driving cycle construction: Winnipeg driving cycle. Transp. Sci. **48**, 170–183 (2014). https://doi.org/10.1287/trsc.1120.0447

31. Huertas, J., Giraldo, M., Quirama, L., Díaz, J.: Driving cycles based on fuel consumption. Energies (Basel) **11**, 3064 (2018). https://doi.org/10.3390/en11113064

32. Mongkonlerdmanee, S., Koetniyom, S.: Development of a realistic driving cycle using time series clustering technique for buses: Thailand case study. Eng. J. **23**, 49–65 (2019). https://doi.org/10.4186/ej.2019.23.4.49

33. Abas, M.A., Rajoo, S., Zainal Abidin, S.F.: Development of Malaysian urban drive cycle using vehicle and engine parameters. Transp. Res. D Transp. Environ. **63**, 388–403 (2018). https://doi.org/10.1016/j.trd.2018.05.015

34. Knez, M., Muneer, T., Jereb, B., Cullinane, K.: The estimation of a driving cycle for Celje and a comparison to other European cities. Sustain. Cities Soc. **11**, 56–60 (2014). https://doi.org/10.1016/j.scs.2013.11.010
35. Tong, H.Y., Ng, K.: Development of bus driving cycles using a cost effective data collection approach. Sustain. Cities Soc. **69**, 102854 (2021). https://doi.org/10.1016/j.scs.2021.102854
36. Cui, Y., Zou, F., Xu, H., Chen, Z., Gong, K.: A novel optimization-based method to develop representative driving cycle in various driving conditions. Energy **247**, 123455 (2022). https://doi.org/10.1016/j.energy.2022.123455
37. Mafi, S., Kakaee, A., Mashadi, B., Moosavian, A., Abdolmaleki, S., Rezaei, M.: Developing local driving cycle for accurate vehicular CO_2 monitoring: a case study of Tehran. J. Clean. Prod. **336**, 130176 (2022). https://doi.org/10.1016/j.jclepro.2021.130176
38. Zhao, X., Zhao, X., Yu, Q., Ye, Y., Yu, M.: Development of a representative urban driving cycle construction methodology for electric vehicles: a case study in Xi'an. Transp. Res. D Transp. Environ. **81**, 102279 (2020). https://doi.org/10.1016/j.trd.2020.102279
39. Qin, X., Yu, K., Li, H., Dai, F., Liu, H., Yang, H., Ye, J., Zhu, H.: Development of a one-day driving cycle for electric ride-hailing vehicles. Transp. Res. D Transp. Environ. **89**, 102597 (2020). https://doi.org/10.1016/j.trd.2020.102597
40. Bhatti, A.H.U., Kazmi, S.A.A., Tariq, A., Ali, G.: Development and analysis of electric vehicle driving cycle for hilly urban areas. Transp. Res. D Transp. Environ. **99**, 103025 (2021). https://doi.org/10.1016/j.trd.2021.103025
41. Tong, H.Y., Ng, K.W.: Developing electric bus driving cycles with significant road gradient changes: a case study in Hong Kong. Sustain. Cities Soc. **98**, 104819 (2023). https://doi.org/10.1016/j.scs.2023.104819
42. Huzayyin, O.A., Salem, H., Hassan, M.A.: A representative urban driving cycle for passenger vehicles to estimate fuel consumption and emission rates under real-world driving conditions. Urban Clim. **36**, 100810 (2021). https://doi.org/10.1016/j.uclim.2021.100810
43. Jia, X., Wang, H., Xu, L., Wang, Q., Li, H., Hu, Z., Li, J., Ouyang, M.: Constructing representative driving cycle for heavy duty vehicle based on Markov chain method considering road slope. Energy AI **6**, 100115 (2021). https://doi.org/10.1016/j.egyai.2021.100115
44. Salihu, F., Demir, Y.K.: Driving cycle for passenger cars on urban roads in Pristina, Kosovo. Baltic J. Road Bridge Eng. **18**, 69–93 (2023). https://doi.org/10.7250/bjrbe.2023-18.589
45. Vámosi, A., Czégé, L., Kocsis, I.: Development of bus driving cycle for Debrecen on the basis of real-traffic data. Period. Polytech. Transp. Eng. **50**, 184–190 (2022). https://doi.org/10.3311/PPtr.16109
46. Chen, Z., Fang, Z., Zhang, Q., Zhou, N., Yu, Q.: Constructing the real-world driving cycle for electric vehicle applications: a comparative study. Trans. Inst. Meas. Control **2022**, 014233122210943. https://doi.org/10.1177/01423312221094384
47. Wang, L., Ma, J., Zhao, X., Li, X.: Development of a typical urban driving cycle for battery electric vehicles based on Kernel principal component analysis and Random Forest. IEEE Access **9**, 15053–15065 (2021). https://doi.org/10.1109/ACCESS.2021.3052820
48. Zhou, S., Jin, J., Wei, Y.: A driving cycle for a fuel cell logistics vehicle on a fixed route: case of the Guangdong province. World Electr. Veh. J. **12**, 5 (2021). https://doi.org/10.3390/wevj12010005
49. Peng, J., Jiang, J., Ding, F., Tan, H.: Development of driving cycle construction for hybrid electric bus: a case study in Zhengzhou, China. Sustainability **12**, 7188 (2020). https://doi.org/10.3390/su12177188
50. Lin, J., Liu, B., Zhang, L.: Autoencoder-based optimization method for driving cycle construction: a case study in Fuzhou, China. J. Ambient. Intell. Humaniz. Comput. **14**, 12635–12650 (2023). https://doi.org/10.1007/s12652-022-04317-7

51. Yuan, M., Kan, X., Chi, C.-H., Cao, L., Shu, H., Fan, Y., Yao, W.: Study of driving cycle of city tour bus based on coupled GA-K-means and HMM algorithms: a case study in Beijing. IEEE Access **9**, 20331–20345 (2021). https://doi.org/10.1109/ACCESS.2021.3054118

52. Zhang, L., Huang, Z., Yu, F., Liao, S., Luo, H., Zhong, Z., Zhu, M., Li, Z., Cui, X., Yan, M., et al.: Road type-based driving cycle development and application to estimate vehicle emissions for passenger cars in Guangzhou. Atmos. Pollut. Res. **12**, 101138 (2021). https://doi.org/10.1016/j.apr.2021.101138

53. Kaymaz, H., Korkmaz, H., Erdal, H.: Development of a driving cycle for Istanbul bus rapid transit based on real-world data using stratified sampling method. Transp. Res. D Transp. Environ. **75**, 123–135 (2019). https://doi.org/10.1016/j.trd.2019.08.023

54. Saleh, W., Kumar, R., Kirby, H., Kumar, P.: Real world driving cycle for motorcycles in Edinburgh. Transp. Res. D Transp. Environ. **14**, 326–333 (2009). https://doi.org/10.1016/j.trd.2009.03.003

55. Seers, P., Nachin, G., Glaus, M.: Development of two driving cycles for utility vehicles. Transp. Res. D Transp. Environ. **41**, 377–385 (2015). https://doi.org/10.1016/j.trd.2015.10.013

José I. Huertas, Jessica Gissella Maradey Lázaro,
and Jenny Díaz Ramírez

Abstract

This chapter focuses on defining driving patterns and exploring their key applications. Additionally, we present various methods to describe these patterns, including driving cycles, while introducing common terminology. The process of experimentally obtaining these patterns is addressed in a separate chapter.

3.1 Driving Patterns

There is not a well-accepted definition for driving patterns [1]. Erroneously, terms such as "driving patterns", "driving cycles", or "characteristic parameters" are sometimes used indistinctly or referred to the same [2–4]. As in [5], some authors have attempted to distinguish them. The concept of driving patterns is often given in terms of its purpose, such as describing the driver's behavior [6–8] or providing the variation of the vehicle speed concerning the duration of a trip [2]; or they could vary depending on the specific region [9–12]. Also, it is said that they are determined by factors such as fleet composition, road network topography, the level of service of the roads, environmental conditions, and

J. I. Huertas (✉) · J. Díaz Ramírez
Sustainable Energy Research Group, Tecnologico de Monterrey, Monterrey 64849, México
e-mail: jhuertas@tec.mx

J. Díaz Ramírez
e-mail: jdiaz@tec.mx

J. G. Maradey Lázaro
Universidad Autónoma de Bucaramanga, 680003 Bucaramanga, Colombia
e-mail: jmaradey@unab.edu.co

driving behavior [13–20], as well as traffic conditions, driving style, and operations modes [12, 21, 22]. In this book, we adopt the most common definition of driving patterns. Consistent to [1, 3, 5, 15, 22], "driving patterns describe the way people drive in a given context (i.e., region, fleet, or route)." Even though this definition is not exact, it agrees with common sense and is helpful.

Driving patterns are frequently used for:

• Assess the energy and environmental performance of vehicles,
• Optimize powertrains and components design and optimization (e.g., battery capacity and electric motor size),
• Formulate vehicle energy management strategies,
• Design energy charging infrastructures,
• Implement incentives for drivers participating in eco-driving programs, and
• Develop intelligent mobility systems.

A confident identification of driving patterns is important, especially for electric vehicles, where driving conditions are essential for energy consumption estimation, autonomy range estimation, and environmental impact quantification [6, 22–27].

Figure 3.1 shows the factors influencing driving patterns, which are classified into three types: driving conditions (external factors), driving style (human factors), and vehicle technology (Technology factors). We highlight that driving style is related to the human decisions that regularly (habits) influence drivers' driving patterns. Therefore, driving style is different from driving patterns.

The main challenge of using driving patterns for the applications listed above is the cross-sensitivity and synergetic effects of the external, human, and technological factors on driving patterns. Fleet managers look after methodologies to decouple human factors from the other influencing factors to evaluate drivers' performance within eco-driving

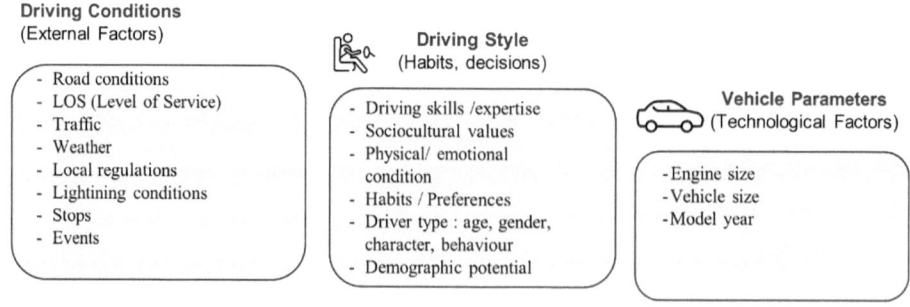

Fig. 3.1 Factors influencing driving patterns

programs. Automakers and government institutions look for methods to decouple the technological aspects from the different factors influencing driving patterns to assess the real performance of their vehicles.

There are several alternatives to describe driving patterns. Next, we will explain the most frequently used.

3.2   Description of Driving Pattern Through Driving Modes

Driving patterns can be described by the relative time the vehicles are operated in a set of driving modes. The most frequent modes considered in the literature are the following ones [9, 28]. The reference values provided within each mode may be different according to the information source:

- Idle: a continuous process during which the engine runs, but the vehicle speed is zero.
- Acceleration: a continuous process when the vehicle acceleration is greater than 0.15 m/s^2.
- Cruise: a continuous process during which the absolute vehicle acceleration or deacceleration is less than 0.15 m/s^2, but the vehicle speed is not zero.
- Deceleration: a continuous process during which the vehicle acceleration is less than −0.15 m/s^2.
- Creeping: a continuous process during which the vehicle speed is between 0 and 5 km/h, not counting zero (i.e., $v \in (0, 5]$ km/h).

Similarly, it is also possible to propose driving modes according to speed ranges, e.g., low-speed zone ($v < 20$ km/h), low-medium ($20 < v < 40$ km/h), medium ($40 < v < 60$ km/h), medium–high ($60 < v < 80$ km/h), and high-speed zone ($v > 80$ km/h); as well as operational modes, as shown in (Fig. 3.2).

3.3   Description of Driving Patterns via SAPDs

A Speed-Acceleration Frequency Distribution (SAFD, Fig. 3.3a) can also describe driving patterns. A SAFD shows the total time spent by the vehicle in each bin of speed and acceleration. When this frequency is normalized and expressed as a percentage of time, it becomes the Speed-Acceleration Probability Distribution (SAPD). This alternative can be considered a direct extension of the driving modes described in the previous section, where each bin of speed acceleration is an operational mode.

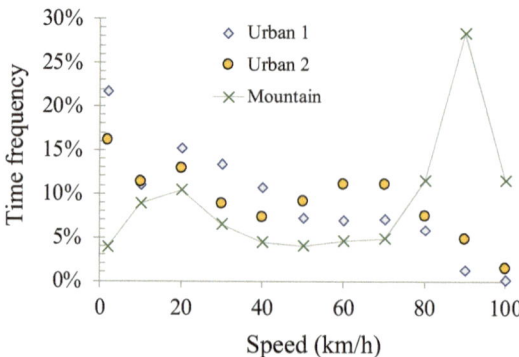

Fig. 3.2 Illustration of the description of driving patterns through operational modes. This plot corresponds to the driving patterns of the fleet of 15 buses serving the route Mexico City-Toluca in a non-stop service. A driving pattern labeled as Urban 1 was obtained with buses moving inside Mexico City, Urban 2 with the same buses operating inside Toluca City, and Mountain corresponds to the driving pattern observed when the identical vehicles traveled between the two cities using a 4 lanes highway [29].

Besides describing driving patterns, SAPDs have been used to decouple human factors from the other factors influencing driving patterns. i.e., SAPDs have been used to identify driving styles. Figure 3.3b highlights the regions with aggressive accelerations and decelerations. The aggressiveness of a driver can be judged by the total fraction of time the vehicle remains in the areas considered aggressive. The limits for aggressiveness can be obtained by observing a large sample of drivers and identifying the 75% percentile of accelerations or deceleration for a given speed. Alternatively, it could be established as the bins with the top 75% fuel consumption for a given speed.

At first glance, we could think that the speed-accelerations bins are directly connected to the operational bins of the engine. i.e., the engine torque (τ_e) versus the engine angular speed (RPM). This assumption arises from the fact that:

$$s = RPM \frac{2\pi}{60} \frac{1}{N_{TD}} \tag{3.1}$$

$$a = \frac{1}{m m_f} \left(\frac{\tau_e N_{TD} \eta_m}{R} - F_x \right) \tag{3.2}$$

where:

s Vehicle speed
RPM Engine angular speed expressed in revolutions per minute
τ_e Engine torque
m Total vehicle weight

Fig. 3.3 Description of driving patterns using SAPD and its connection with the engine torque versus RPM operational modes of the engine. **a** SAPD obtained by observing the operational working conditions of 15 buses serving the route Mexico City-Toluca in a non-stop service. **b** The previous SAPD as observed on the engine torque versus RPM diagram and the observed specific fuel consumption at each bin of τ_e-RPM [29]

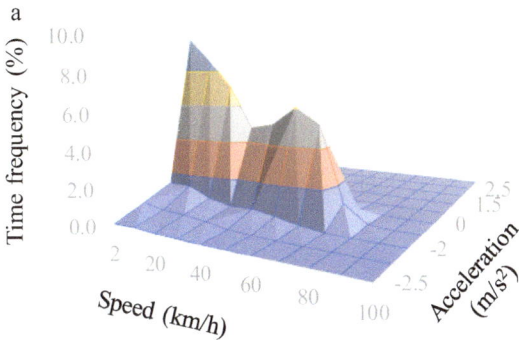

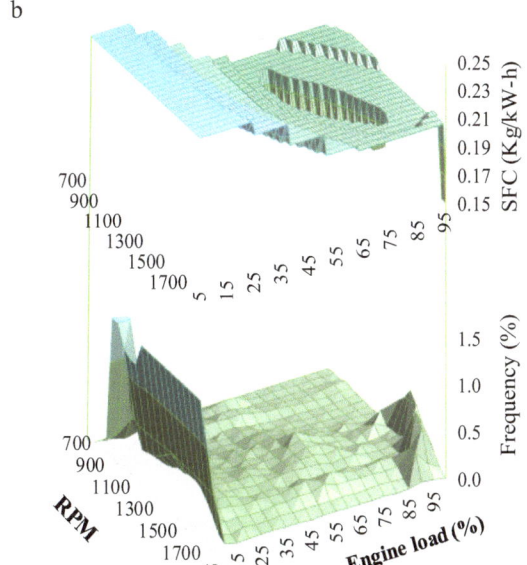

m_f	Mass fraction
η_m	mechanical efficiency of the transmission
R	Wheel radius
F_x	Traction force at the wheel needed to overcome the road loads.

There is a possibility that the speed-acceleration bins of the SAPD have a direct connection to the RPM-τ_e operational bins. This is very attractive because, in that case, the SAPD directly describes the loads to the engine. However, the correspondence between s and RPM depends on the transmission ratio (N_{TD}), which varies during the vehicle's operation, breaking the one-to-one relation. Similarly, it happens with acceleration (a) and engine torque (τ_e). However, this connection can be leveraged to decouple the technology factors from the other factors influencing driving patterns, which will be discussed later in Chap. 6.

3.4   Description of Driving Patterns Through VSP

Vehicle Specific Power (VSP) is the ratio of the power demanded to the engine and the vehicle's total mass (m). Some authors use the power demanded to the vehicle wheels instead of the power demanded to the engine in the VSP definition. The difference between them is the energy losses in the power transmission from the engine to the wheels through the vehicle power train. Even though this difference could be considered negligible, it significantly affects vehicle energy consumption through inertial forces, as explained later. Thus, in this book, we prefer the definition associated with the power demanded by the engine. Figure 3.4 shows a typical VSP frequency distribution obtained during vehicle operation monitoring.

Next, we will describe a model to estimate the power demanded to the engine (P_e) of ground vehicles. This model is well known and has become a key concept directly related to evaluating the energy performance of vehicles. We recommend readers to master this model, which can be found in any textbook related to vehicle dynamics [30]. It is an energy balance within the vehicle.

The forces that restrict the free movement of the vehicle (Fig. 3.5) are the rolling resistance (R_x. Equation 3.3), drag (F_d, Eq. 3.4), and gravity (F_g, Eq. 3.5). The traction force at the wheel (F_x, Eq. 3.6) overcomes those forces and the force required to accelerate the vehicle (the translational inertial forces). This last force includes the total mass of the vehicle (rotating and non-rotating masses).

$$R_x = f_r mg \cos \theta \qquad\qquad (3.3)$$

Fig. 3.4 Typical VSP frequency distribution observed in the everyday operation of vehicles

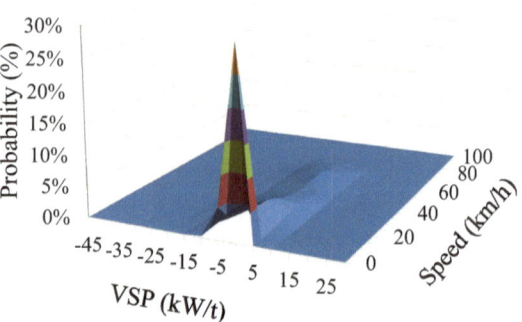

Fig. 3.5 Forces acting on vehicles that oppose the motion of vehicles moving up a road with a positive grade. Taken from [31]

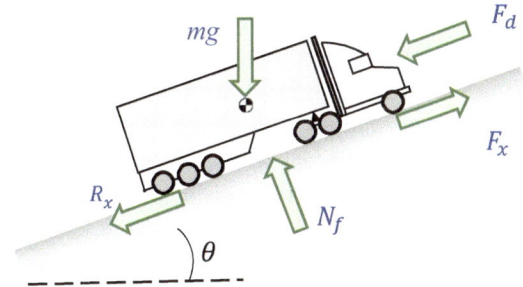

$$F_d = \frac{1}{2} c_d A_f \rho_{air} V^2 \tag{3.4}$$

$$F_g = mg \sin \theta \tag{3.5}$$

$$F_x - R_x - F_d - F_g = ma \tag{3.6}$$

$$F_x = \left(\frac{\tau_e N_{TD_i} \eta_m}{R} - \frac{I_i}{R^2} a \right) \tag{3.7}$$

$$P_e = \tau_e \frac{V}{R N_{TD_i}} \tag{3.8}$$

$$m_{f_i} = 1 + 0.04 (N_{TD_i}) + 0.0025 (N_{TD_i})^2 \tag{3.9}$$

$$P_e = \frac{\left(mm_{f_i} a + f_r Mg \cos \theta + \frac{1}{2} c_d A_f \rho_{air} V^2 + mg \sin \theta \right) V}{\eta_T} = \dot{v}_f \, \rho_f LHV \, \eta_{th} \eta_e \tag{3.10}$$

$$VSP = \frac{\left(mm_{f_i} a + f_r Mg \cos \theta + \frac{1}{2} c_d A_f \rho_{air} V^2 + mg \sin \theta \right) V / \eta_T}{m} \tag{3.11}$$

The engine provides the tractive force required at the wheel (Eq. 3.7) through the vehicle power train that operates at a given transmission ratio (N_{TD_i}). The engine must provide the torque required at the wheel ($\tau_e N_{TD_i} \eta_T$) plus the rotating inertia forces associated with all rotating components ($\frac{I_i}{R} a$), including the internal parts of the engine (I_i). Thus, this tractive force is equal to the torque perceived at the wheel divided by the vehicle wheel radius minus the equivalent inertia forces of the rotating power-train component, as shown in Eq. 3.7.

Comparing Eqs. 3.6 and 3.7, we can obtain an expression for the engine torque (τ_e). This expression (not shown) should be multiplied by the engine rotational speed (ω_e) to obtain an expression for the minimum engine power required to move the vehicle ($P_e = \tau_e \omega_e$). The resulting expression is Eq. 3.10. In this equation, the rotational inertial force (F_i) is expressed in terms of an equivalent mass (m_e) in the way that $F_i = m_e a$. Thus $m_e = \frac{I_i}{R}$. Several authors have combined the translational and rotational masses by using the means of a mass fraction (m_f) such as $m + m_e = m m_f$. Equation 3.9 is the experimental expression used most of the time for m_f.

The power delivered by the engine is provided by an external energy source (fuel or battery). When the source is a fuel, the fuel consumption rate $(\dot{v}_f)$ can be computed from $P_e = \rho_f \dot{v}_f LHV \eta_{th} \eta_e$. The vehicle's tank-to-wheel energy efficiency is the product of the mechanical engine efficiency (η_e), the engine's thermal efficiency (η_{th}), and the mechanical transmission efficiency (η_t).

Finally, the instant VSP (Eq. 3.11) is obtained by dividing Eq. 3.10 by the vehicle's total mass. Notice that VSP is directly linked to vehicle fuel/energy consumption. Therefore, the main advantage of describing driving patterns through frequency distributions of VSPs is that VPS connects driving patterns with vehicle fuel consumption and, by extension, to tailpipe emissions [29].

3.5 Description of Driving Patterns Through Characteristic Parameters

Driving patterns can be described by a set of characteristic parameters (CPs) [1, 6, 10, 12, 21, 32–35]. There is not a clear definition for CPs. They are metrics based on speed and time, like average speed, average positive acceleration, percentage of idle time, etc. [35]. Some of the CPs naturally describe the way people drive. For example, it is intuitive to say that in a given region, people drive at a given average speed, and therefore, average speed is a relevant CP to describe driving patterns. However, the average intensive kinetic energy $(V^2/2)$ is not an intuitive but well-accepted CP since it is connected to vehicle fuel consumption.

No consensus exists on which CPs should be used or how many are enough to describe a driving pattern fully. Table 3.1 lists CPs reported in the literature [5, 6, 36]. However, some important issues regarding CPs should be considered.

CPs refer to how people drive a vehicle. Therefore, parameters that describe a single part of the vehicle, such as the engine or battery, are not CPs. The same situation happens with the road. The road grade or road conditions are not CPs. All these parameters influence the driving pattern and should be described and considered, but they do not describe how people drive.

Some other authors extended the previous definition and included the average specific fuel consumption (L/100 km) or battery usage [37, 38] and emission indexes (EI) (g/km) of tailpipe pollutants [34] as CPs. In this book, we prefer to preserve the above definition and exclude SFC and EI from the family of acceptable CPs.

CPs always refer to metrics that describe a population. Thus, the average speed is a CP, but the instant speed is not a CP. A similar issue happens with maximum speed. The

Table 3.1 Characteristic parameters (CPs) used in the literature to describe driving patterns

Speed	Spatial	Vehicle dynamics
Average trip speed v	Total (trip) distance d length of driving period	Positive kinematic energy (PKE)
Average driving speed v_x	Daily driving distance	Root mean square of acceleration (RMSA)
Standard deviation of speed σ_v	Slope θ	Vehicle-specific power (VSP)
Maximum speed v_{max}	Positive slope θ^+	Maximum VSP
95th percentile of speed v_{95}	Negative slope θ^-	Minimum VSP
Other functions of speed: $v\sin\theta, v^3, a^+v$	Average slope	Average positive VSP
		Average negative VSP
		Relative positive acceleration (RPA)
		Relative positive speed []
		Relative real speed (RRS) []
		Relative cubic (positive) speed
		Relative square speed (RSS)

Time	**Acceleration**	**Battery**
Total time T	Average acceleration a	Torque τ
Acceleration time (t^{a^+})	Maximum acceleration a_{max}	State of charge (SoC)
Deceleration time (t^{a^-})	Average positive acceleration a^+	Battery current e
Driving time t	Average deceleration a^-	Battery power P
Idle time t_i	Number of a^+/a^- changes	
Creeping time t_{crp}	Number of accelerations/km	
Cruising time t_c	Standard deviation of positive acceleration σ_{a^+}	**Stops**
Time % of acceleration mode	Standard deviation of negative acceleration σ_{a^-}	Number of stops
Time % of creeping mode	Interquartile range of speed	Stops per km
Time % of cruising mode	95th percentile of a^+ : a_{95}^+	**Emissions**
Time % of idle mode	95th percentile of a^- : a_{95}^-	NO$_x$	CO$_2$
Standing time	Relative a^+	CO	HC

maximum individual speed of vehicles should not be considered a CP, but the average maximum speed could be a valid CP. Similarly, this happens with idling time.

CPs do not depend on monitoring aspects. Thus, the total distance traveled by all vehicles monitored is not a CP. It grades the duration of the monitoring campaign. However, the average daily distance traveled per person is a CP, which is seldom considered in the literature. Similarly, this happens with idling time. The total or average idling time is not a CP because it depends on the trip duration. However, the average percentage of idling time is a valid CP.

Most CPs are single values, but the frequency distribution of that parameter could be more meaningful. For example, the average VSP is not a valid CP, but the frequency distribution of VSP is a valid alternative for describing driving patterns.

Some average values are naturally zero. For example, the average acceleration is not a valid CP because it is always zero. The average positive or negative acceleration or the standard deviation of the acceleration are valid CPs. In this book, we use positive and negative acceleration (instead of acceleration or deceleration) to emphasize this aspect.

Figure 3.6 shows the most frequent CPs used in the literature to describe driving patterns. It shows that average speed, average positive acceleration, and percentage of idling time are the most common CPs. Most studies use between 6 and 15 CPs, with an average of 10 CPs, to describe driving patterns.

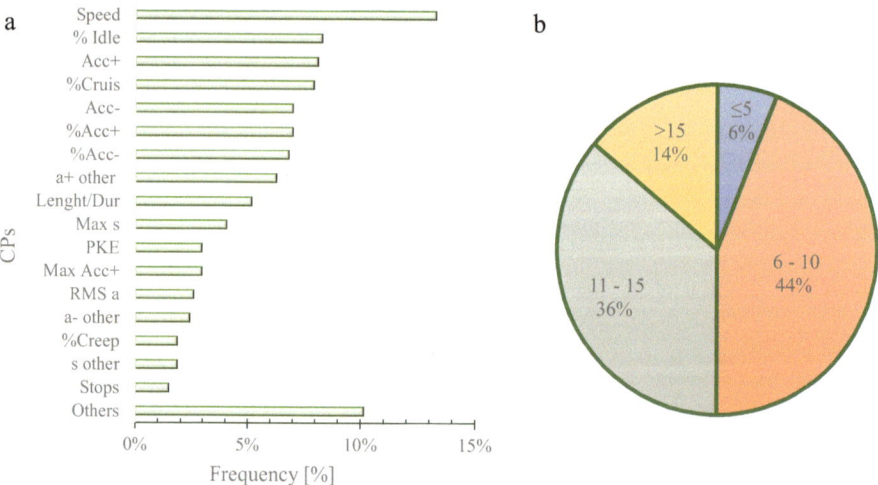

Fig. 3.6 Characteristic parameters used to represent driving patterns. **a** Most frequently used CPs. **b** Number of CPs used

3.6   Description of Driving Patterns Through Driving Cycles

Regarding the definition of a driving cycle (DC), the research community is more in agreement: A driving cycle is a time series of speed that represents the driving patterns in a specific region [9, 12, 13, 15–18, 23, 32, 39–49]. Sometimes, they are developed for a particular vehicle technology or an application of interest, for example, for bus rapid transit (BRT) systems [21, 47, 50]. Figure 3.7 shows a DC illustration.

DCs are mainly used for the following purposes:

- Assessment of the vehicle's fuel/energy consumption, power, and energy efficiency [2, 17, 24, 35, 43, 44, 51–54].
- Determination of vehicle tailpipe emissions [18, 20, 33, 35, 48, 55–59].
- Development of procedures for designing, testing, and homologating any type of vehicle [5, 42].
- Design of vehicle powertrains, energy storage, and control systems in vehicles, especially in electric and hybrid vehicles [7, 20, 42, 49, 50, 52, 54, 55, 60, 61].
- Certification of the fuel economy and driving ranges of vehicle technologies [40, 60];
- Improve vehicle operation [23, 51].
- Simulate and design traffic control and city management systems [24, 40].
- Evaluate operative costs of emergent vehicle technologies.

3.6.1   Properties of Driving Cycles

Next, we will describe some properties that a DC must satisfy.

Representativeness
This concept refers to the DC's ability to represent driving patterns. A DC represents a driving pattern when its characteristics parameters (CPs*) are similar to those of the

Fig. 3.7 Illustration of driving cycles and frequent terminology used when dealing with driving cycles

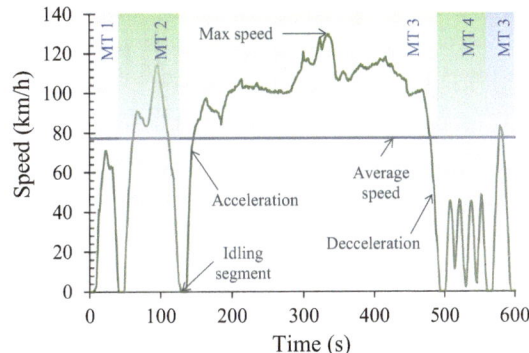

driving patterns (CPs). Note that the characteristics parameters of the DC are labeled with a*, while the characteristics parameters of the driving pattern do not have a (*). Some authors name the characteristic parameters of the DC as performance values.

In this way, a representative DC should have the same average speed, percentage of idle time, and average positive acceleration as the driving pattern. Therefore, the degree of representativeness can be assessed through Eq. 3.12. RD_i is the relative difference between the CP_i* that describes the DC and the CP_i that describes the driving pattern. i denotes any of the CPs used as assessment parameters. RD_i could vary between zero and infinity, but the criteria of representativeness demand that the DC has $RD_i < 5\%$. Some authors have relaxed this threshold value up to 15% or even 20% [36, 39, 45, 62, 63].

$$RD_i = \frac{\left|CP_i^* - CP_i\right|}{CP_i} \tag{3.12}$$

A large discrepancy exists in the number of CPs that must be used to evaluate representativeness. The tendency is to include as many CPs as possible. Furthermore, some authors have included fuel (energy) consumption and tailpipe emissions [64, 65]. In this way, a DC represents a driving pattern when a given technology exhibits the same fuel consumption following the DC on a chassis dynamometer as observed in the daily use of the same technology in the region of interest. We highly recommend including the energy consumption criteria within the definition of representativeness, and when the upcoming technology allows the measurement of the daily tailpipe emissions, include tailpipe emissions within the criteria of representativeness.

Within the community of the DCs, it is implicit that when someone refers to a DC, he/she means a representative DC. Otherwise, the speed vs. time series is named a trip. However, the lack of explicitness has caused some studies to skip the representativeness validation of their potential DCs [11, 21, 28]. We recommend keeping DC exclusively for representative driving cycles.

Finally, some authors have stated that the representativeness of a DC depends on the quantity and quality of data collected during the monitoring campaign of the operation of the vehicles [5, 15, 20, 28, 33, 41, 48, 62, 66] i.e., it depends on the quantity and quality of the data obtained to grasp the driving pattern. The quantity refers to the number of vehicles monitored and the duration of the monitoring campaign. The quality refers to the data quality analysis carried out to ensure the absence of erroneous data. Some other authors have pointed out that the representativeness of the DC also depends on the objectiveness in selecting the routes [45, 67]. These comments underline the relevance of correctly grasping the driven pattern as a prior step to constructing a DC that represents it. Otherwise, obtaining a DC that represents the input data is possible, but that input data may not accurately capture the driving pattern. Section 4.5 will address the issues related to the sample size needed to assure representativeness in grasping the driving pattern.

Reproducibility

DC must be easily followed by the vehicle technology that they were meant to be created for. A regular driver using any vehicle technology within the segment of the objective vehicles should be able to reproduce the cycle on the road or a chassis dynamometer without excessive troubles. This requirement imposes restrictions on continuity, time resolution, and smoothness.

It is well-accepted that the time resolution of the DC is 1 s (1 Hz). Higher frequencies are irrelevant for drivers (They cannot follow the cycle), and lower frequencies contribute to the variability of the results obtained in the tests where the DC is used. The continuity condition imposes that a speed must be specified at any time within the DC. The condition of smoothness prevents the presence of abrupt or unrealistic accelerations (speed time derivative). At any time, the acceleration must be within values that the engine can provide to the vehicle.

Duration

Since DCs are used as part of tests that usually are time consuming, require well-trained personnel, and use expensive consumables, technicians and researchers look for short DCs. However, DCs that are too short are associated with large uncertainties in the results obtained. These tests include many sources of errors that are averaged out with the duration of the DC. Researchers have found that about 20 min is an acceptable tradeoff between cost and results precision. Section 5.6 further discusses the issues related to the duration of DC.

Uniqueness

Identifying the DC that best represents the local driving pattern under study is highly desirable. However, as shown in Chap. 5, the methods for constructing DC fail to construct a unique DC that achieves this objective. They construct many DCs with similar degrees of representativeness of the same driving pattern [21, 39, 46, 48, 55]. Therefore, the DC obtained for a city or a given application usually undergoes a socialization process during which the stakeholders adopt a proposed DC as "the DC" for the region of interest. The lack of uniqueness issue worsens when people consider that driving patterns change with time as urban mobility improvements occur or the lack of actions increases traffic congestion.

Topography

DCs for cities located in non-flat regions are unknown [68, 69]. There is a need for DC construction methods that include the effect of road grade. Some authors have attempted to address this problem by accompanying the DC with a time series of road grades to include their effect on fuel consumption during the tests. However, questions about how this action contributes to increasing the degree of representativeness of driving patterns in non-flat regions have not been addressed. Section 5.5 will suggest alternatives to construct DCs for non-flat regions.

3.6.2 DC Classification

DCs can be classified according to their application or characteristics [70] (Fig. 3.8).

Regarding their application, DCs are classified into legislative and non-legislative. Legislative DCs are those DCs that stakeholders, especially governmental authorities, have adopted as official DCs. These DCs are also known as homologation DCs because manufacturers use them to certify the compliance of their vehicles with the local regulations regarding specific fuel consumption and emission indexes [39]. The homologation DCs do not necessarily represent the local driving cycles. In fact, they very seldom were obtained from a process of monitoring the actual region pattern. The non-legislative DCs are all those DCs obtained by researchers monitoring the real operating conditions of vehicles. Manufacturers use them to improve the energy efficiency of their vehicles, and by governmental authorities to estimate vehicular emissions. These DCs must be adopted as legislative DCs before being used for vehicle certification.

Regarding the DC characteristics, DCs can be classified as modal or transient. Modal DCs are artificially synthesized DCs that look for testing the vehicle under different modes of operation, such as idling, constant acceleration, cruise speed, and constant deceleration (Fig. 3.7). They are easy to reproduce on a chassis dynamometer or road tests. However, due to the nature of these cycles, the transitions between various modes are somewhat artificial in nature. It is under controversy their use for evaluating fuel consumption (FC)

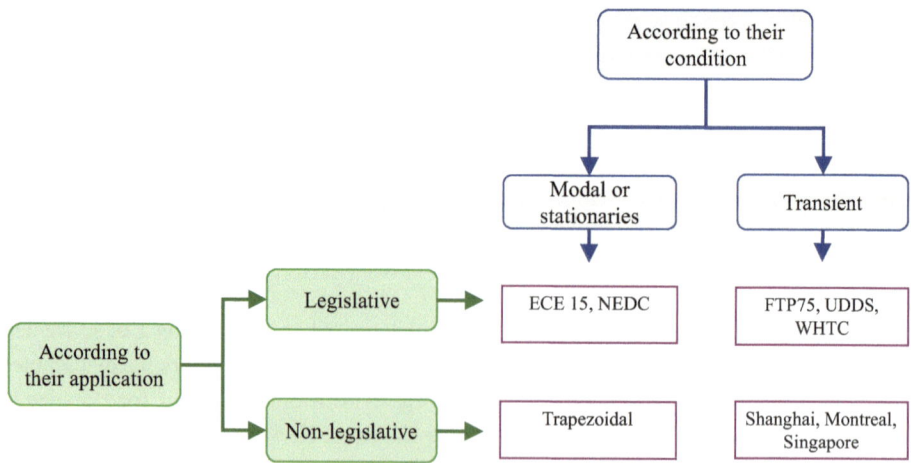

Fig. 3.8 Classification of driving cycles. Based on [10, 12, 19, 51, 70]

[10] and tailpipe emissions [12, 55]. Transient or real DCs are time series constructed as the collection of segments of trips obtained from vehicles under real operation conditions. These DCs guarantee their reproducibility on a chassis dynamometer or road tests.

3.6.3 Use of DCs for the Determination of Fuel Consumption and Tailpipe Emissions

Considering that the understanding of how DCs are deployed will facilitate the understanding of the different phases involved in their construction, this section describes the use of DC for the most frequent application, which is the determination of fuel consumption and tailpipe emissions.

Tests on Chassis Dynamometers

The constant volume sampling (CVS) system is the worldwide well accepted procedure to measure vehicle tailpipe emissions. In it, the vehicle is secured on a chassis dynamometer (Fig. 3.9), which is adjusted to simulate the road rolling resistance, the aerodynamic vehicle resistance, and the inertial forces [71]. The fuel consumption and tailpipe emission are measured while a trained driver forces the vehicle to reproduce the DC of interest [72].

Tailpipe emissions are collected and diluted with filtered fresh air and pumped out of the laboratory at a constant volume rate. A powerful blower is used to avoid the accumulation of vehicle emissions within the system, and an online venturi tube working under stagnant conditions ensures the extraction of the diluted gases at a constant volumetric rate. Using the same principle, a small sample of the diluted vehicular emissions is collected at a constant rate in non-reacting sieve bags using the same extraction principle mentioned before. When the DC is completed, each pollutant concentration is measured following well-accepted procedures (non-dispersed infrared for OC and CO_2 and non-dispersive ultraviolet for NO and NO_2). The resulting concentration is proportional to the total mass of that pollutant emitted during the test. The constant of proportionality of the CVS system is obtained through an independent calibration test. Results are reported in terms of emission indexes which typically are expressed in gr/km.

During the same test, the fuel consumption can be measured, simultaneously, by following the so-called gravimetric method. In this case, the lines that supply the fuel from the vehicle tank to the engine are deviated toward an external tank, and the fuel consumption is measured as the difference in weight of the tank after and before the test. Results are reported as specific fuel consumption (L/100 km) or fuel economy (km/L). In these tests, technicians must avoid interfering with the normal vehicle operational conditions when manipulating the fuel lines. They must ensure that the fuel pump used in the test is the same as the vehicle used in its normal operation.

The development of these tests is expensive. The cost of a CVS laboratory is high (2–5 million USD). Carrying out the tests is time-consuming, requires trained personnel,

a

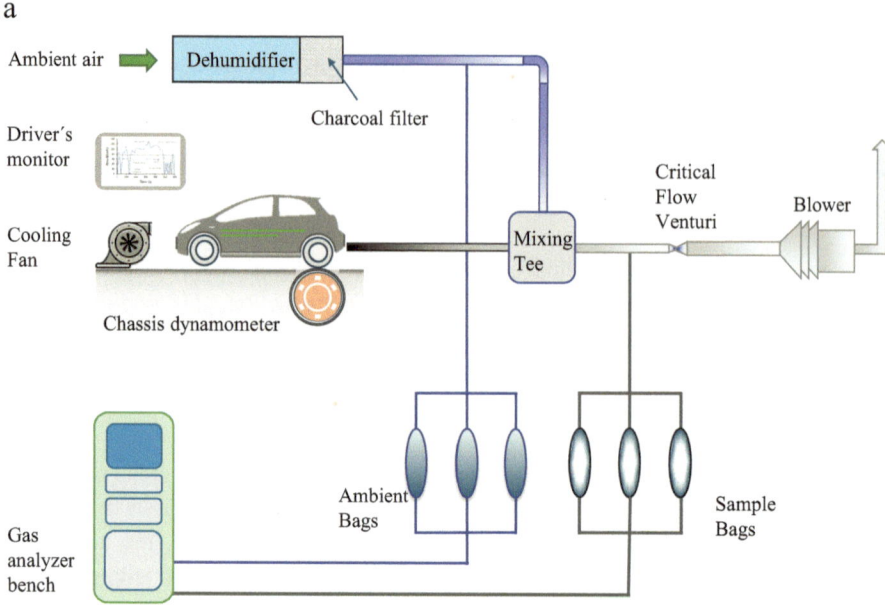

b

Fig. 3.9 Measurement of tailpipe emissions and fuel consumption following a driving cycle on a chassis dynamometer using a CVS. **a** Schematic representation of a chassis dynamometer emissions test facility. **b** Actual picture of a CVS system

and costly consumables. It has been shown that the results obtained on these tests are different from those observed during the normal use of the vehicles. Therefore, with the advance of alternative technologies to measure the real emissions and fuel consumption of vehicles, these tests have become limited for tests conducted for regulatory purposes.

On Road Tests

For non-regulatory purposes, the on-road measurements (ORMs) have become the preferred method for determining tailpipe emissions and vehicle fuel consumption. In this case, the emission is directly measured from the tailpipe by a portable emission monitoring system (PEMS, [73, 74]), while the vehicle reproduces the targeted DC (Fig. 3.10). PEMS were designed to measure in-use emissions during real-world on-road operation under various ambient conditions, traffic, and operational duty cycles. PEMS has become a reliable instrumental system for measuring tailpipe emissions and fuel consumption. Extensive official programs have confirmed the ability of the most advanced PEMS to provide a measurement quality in line with the emissions certification labs [74]. They can be installed on various vehicles, including those that are impractical to test on a chassis dynamometer [75].

PEMS can determine the regulated gaseous pollutant emissions (CO, CO_2, THC, and NOx) in concentration units (% or ppm by volume). Simultaneously, the instantaneous flow of exhaust gases is measured with a Pitot tube or using the intake manifold air mass flowrate signal provided by the OBD. The air–fuel ratio derived from the measured

a   b

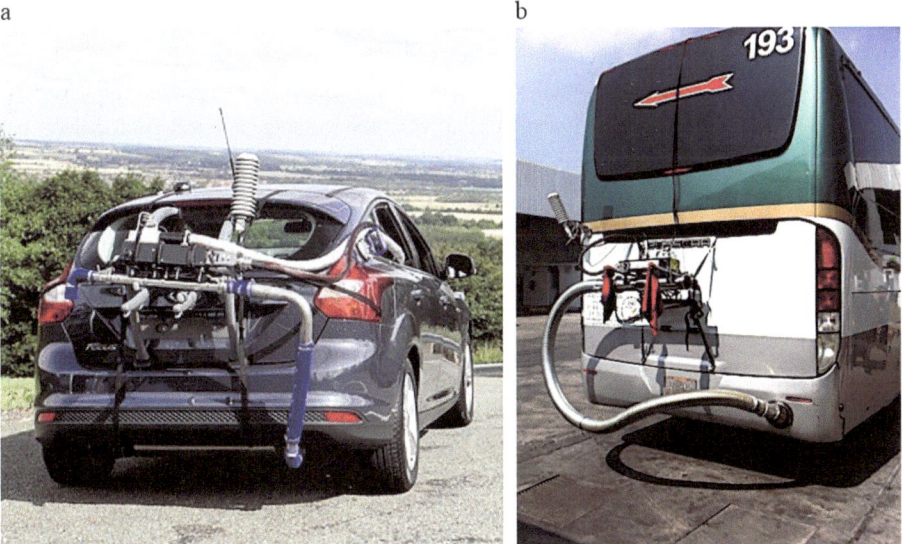

Fig. 3.10 Measurement of tailpipe emissions and fuel consumption following a driving cycle on the road using a PEMS in **a** light-duty vehicles and **b** heavy-duty vehicles

pollutants ratios made it possible to determine the mass flow rate. Results are reported in terms of mass emitted per kilometer traveled (gr/km). An electric low-pressure impactor (ELPI) in cascade with a double gas dilution system is used to determine the mass flow rate of particulate matter (PM). A real-time engine exhaust particle spectrometer (EEPS) can also be used for this purpose [73]. This system provides PM emissions both in mass (grams of PM/km) and in number of particles (PN) emitted (#/km). Signal correction and synchronization should be conducted before obtaining the mass rate of emission index report.

Fuel or energy consumption measurements are conducted following the same protocol described previously and specified in SAE J1321-202010. It can also be obtained directly from the engine computer unit (ECU) via the OBD system [71].

On-road tests require meticulous planning due to the need to coordinate various simultaneous activities, ensure the safety of the many technicians involved, and ensure the correct operation of the expensive equipment used during the tests.

In the planning stage, it is critical to design efficient validation methods to ensure the relevance and usefulness of the data collected. The selection of road tracks and schedules for the tests is vital, considering factors such as topographic characteristics and traffic that may influence the execution of the tests and the results obtained.

The execution phase begins with the vehicle preparation. It includes the payload simulation, which usually is carried out by water tanks that simulate passengers in buses or sandbags that simulate the payload in freight vehicles. It also includes the proper installation of the instrumentation adapted to the unique characteristics of the tested vehicle to ensure the equipment's safety and eliminate or mitigate mechanical vibration effects and electric noise from the energy power supply. Then, the correct operation of the data gathering system and dataloggers should be verified. Calibration of sensors must be carried out following manufacturer recommendations. It is important to reduce the variability originated while the driver becomes familiar with the vehicle and test protocol by performing preliminary tests. Data should be analyzed in real time to identify problems in the operation of the instrumental system. It is also advisable that the sensors be recalibrated at the end of each test to ensure the accuracy of the results. Finally, each test must be checked for repeatability given the variable conditions on the road [76].

ORMs are projected to be a very competitive method, given the improvements in precision and accuracy in emission measurements technologies, the reduction of PEMS' weight and costs, and the reliability of data gathering, data transmission (telemetry), and data treatment methods. However, ORMs currently face important challenges, such as the high associated costs and the complicated logistics involved in the tests.

3.7 Concluding Remarks

This chapter has provided a holistic approach to the concept of driving patterns and driving cycles (DC). Initially, we stated that "driving patterns describe how people drive in a given context (i.e., region, fleet, or route) and highlighted that it is the most accepted definition for driving patterns. Then, we described the different methods of describing driving patterns. It includes a set of characteristic parameters (CPs), the frequency distribution of speed-accelerations (SAFD) or speed-acceleration probability distribution (SAPD); the frequency distribution of vehicle-specific power (VSP); and DCs. This last alternative has enabled the use of driving patterns to evaluate the energy and environmental performance of vehicles under the local conditions of use, which has been used for regulatory purposes, optimization of the vehicle's power train, and optimization of the vehicle's energy management systems. The next chapter will present the existing methodologies for constructing DCs.

References

1. Huertas, J., Giraldo, M., Quirama, L., Díaz, J.: Driving cycles based on fuel consumption. Energies (Basel) **11**, 3064 (2018). https://doi.org/10.3390/en11113064
2. Tamsanya, S., Chungpaibulpatana, S., Limmeechokchai, B.: Development of a driving cycle for the measurement of fuel consumption and exhaust emissions of automobiles in Bangkok during peak periods. Int. J. Automot. Technol. **10**, 251–264 (2009). https://doi.org/10.1007/s12239-009-0030-4
3. Feng, L., Liu, W., Chen, B.: Driving pattern recognition for adaptive hybrid vehicle control. SAE Int. J. Alternat. Powertrains **1**, (2012). https://doi.org/10.4271/2012-01-0742
4. Lin, J., Niemeier, D.A.: An exploratory analysis comparing a stochastic driving cycle to California's regulatory cycle. Atmos. Environ. **36**, 5759–5770 (2002). https://doi.org/10.1016/S1352-2310(02)00695-7
5. Berzi, L., Delogu, M., Pierini, M.: Development of driving cycles for electric vehicles in the context of the city of Florence. Transp. Res. D Transp. Environ. **47**, 299–322 (2016). https://doi.org/10.1016/j.trd.2016.05.010
6. Li, X., Zhang, Q., Peng, Z., Wang, A., Wang, W.: A data-driven two-level clustering model for driving pattern analysis of electric vehicles and a case study. J. Clean. Prod. **206**, 827–837 (2019). https://doi.org/10.1016/j.jclepro.2018.09.184
7. Huertas, J.I., Díaz, J., Giraldo, M., Cordero, D., Tabares, L.M.: Eco-driving by replicating best driving practices. Int. J. Sustain. Transp. **12**, 107–116 (2018). https://doi.org/10.1080/15568318.2017.1334107
8. Ericsson, E.: Independent driving pattern factors and their influence on fuel-use and exhaust emission factors. Transp. Res. D Transp. Environ. **6**, 325–345 (2001). https://doi.org/10.1016/S1361-9209(01)00003-7
9. Yuhui, P., Yuan, Z., Huibao, Y.: Development of a representative driving cycle for urban buses based on the K-means cluster method. Cluster Comput. **22**, 6871–6880 (2019). https://doi.org/10.1007/s10586-017-1673-y

10. Kamble, S.H., Mathew, T.V., Sharma, G.K.: Development of real-world driving cycle: case study of Pune, India. Transp. Res. D Transp. Environ. **14**, 132–140 (2009). https://doi.org/10.1016/j.trd.2008.11.008

11. Fotouhi, A., Montazeri-Gh, M.: Tehran driving cycle development using the K-means clustering method. Scientia Iranica **20**, 286–293 (2013). https://doi.org/10.1016/j.scient.2013.04.001

12. Knez, M., Muneer, T., Jereb, B., Cullinane, K.: The estimation of a driving cycle for Celje and a comparison to other European cities. Sustain. Cities Soc. **11**, 56–60 (2014). https://doi.org/10.1016/j.scs.2013.11.010

13. Zou, S., Zhu, X., Xu, L., Zhu, H.: Construction of vehicle driving cycle in Fuzhou. J. Phys. Conf. Ser. **1453**, 012146 (2020). https://doi.org/10.1088/1742-6596/1453/1/012146

14. Lipar, P., Strnad, I., Česnik, M., Maher, T.: Development of urban driving cycle with GPS data post processing. PROMET - Traffic Transp. **28**, 353–364 (2016). https://doi.org/10.7307/ptt.v28i4.1916

15. Ma, R., He, X., Zheng, Y., Zhou, B., Lu, S., Wu, Y.: Real-world driving cycles and energy consumption informed by large-sized vehicle trajectory data. J. Clean. Prod. **223**, 564–574 (2019). https://doi.org/10.1016/j.jclepro.2019.03.002

16. Achour, H., Olabi, A.G.: Driving cycle developments and their impacts on energy consumption of transportation. J. Clean. Prod. **112**, 1778–1788 (2016). https://doi.org/10.1016/j.jclepro.2015.08.007

17. Geetha, A., Subramani, C.: Development of driving cycle under real world traffic conditions: a case study. Int. J. Electr. Comput. Eng. (IJECE) **9**, 4798 (2019). https://doi.org/10.11591/ijece.v9i6.pp4798-4803

18. Peng, Y., Zhuang, Y., Yang, Y.: A driving cycle construction methodology combining *k*-means clustering and Markov model for urban mixed roads. Proc. Inst. Mech. Eng. Part D: J. Autom. Eng. **234**, 714–724 (2020). https://doi.org/10.1177/0954407019848873

19. Tzirakis, E., Zannikos, F.: Development of processing methodologies used to form complete driving-cycle dynamometer tests based on urban on-road driving and road gradient data. Proc. Inst. Mech. Eng. Part D: J. Autom. Eng. **229**, 97–110 (2015). https://doi.org/10.1177/0954407014529940

20. Nouri, P., Morency, C.: Evaluating microtrip definitions for developing driving cycles. Transp. Res. Record: J. Transp. Res. Board **2627**, 86–92 (2017). https://doi.org/10.3141/2627-10

21. Ho, S.-H., Wong, Y.-D., Chang, V.W.-C.: Developing Singapore driving cycle for passenger cars to estimate fuel consumption and vehicular emissions. Atmos. Environ. **97**, 353–362 (2014). https://doi.org/10.1016/j.atmosenv.2014.08.042

22. Gong, Q., Midlam-Mohler, S., Marano, V., Rizzoni, G.: An iterative Markov chain approach for generating vehicle driving cycles. SAE Int. J. Engines **4**, (2011). https://doi.org/10.4271/2011-01-0880

23. Lai, J., Yu, L., Song, G., Guo, P., Chen, X.: Development of city-specific driving cycles for transit buses based on VSP distributions: case of Beijing. J. Transp. Eng. **139**, 749–757 (2013). https://doi.org/10.1061/(ASCE)TE.1943-5436.0000547

24. Tsai, J.-H., Chiang, H.-L., Hsu, Y.-C., Peng, B.-J., Hung, R.-F.: Development of a local real world driving cycle for motorcycles for emission factor measurements. Atmos. Environ. **39**, 6631–6641 (2005). https://doi.org/10.1016/j.atmosenv.2005.07.040

25. Ashtari, A., Bibeau, E., Shahidinejad, S.: Using large driving record samples and a stochastic approach for real-world driving cycle construction: Winnipeg driving cycle. Transp. Sci. **48**, 170–183 (2014). https://doi.org/10.1287/trsc.1120.0447

26. Wang, J., Huang, Y., Xie, H., Tian, G.: Driving pattern prediction model for hybrid electric buses based on real-world driving data. In: Proceedings of the 28th International Electric Vehicle Symposium and Exhibition (2015)

27. He, H., Sun, C., Zhang, X.: A method for identification of driving patterns in hybrid electric vehicles based on a LVQ neural network. Energies (Basel) **5**, 3363–3380 (2012). https://doi.org/10.3390/en5093363

28. Seers, P., Nachin, G., Glaus, M.: Development of two driving cycles for utility vehicles. Transp. Res. D Transp. Environ. **41**, 377–385 (2015). https://doi.org/10.1016/j.trd.2015.10.013

29. Giraldo, M., Huertas, J.I.: Real emissions, driving patterns and fuel consumption of in-use diesel buses operating at high altitude. Transp. Res. D Transp. Environ. **77**, 21–36 (2019). https://doi.org/10.1016/j.trd.2019.10.004

30. Gillespie, T.: Fundamentals of Vehicle Dynamics. Society of Automotive Engineers, Warrendale (1992)

31. Huertas, J.I., Serrano-Guevara, O., Díaz-Ramírez, J., Prato, D., Tabares, L.: Real vehicle fuel consumption in logistic corridors. Appl. Energy **314** (2022). https://doi.org/10.1016/j.apenergy.2022.118921

32. Wang, Q., Huo, H., He, K., Yao, Z., Zhang, Q.: Characterization of vehicle driving patterns and development of driving cycles in Chinese cities. Transp. Res. D Transp. Environ. **13**, 289–297 (2008). https://doi.org/10.1016/j.trd.2008.03.003

33. Pouresmaeili, M.A., Aghayan, I., Taghizadeh, S.A.: Development of Mashhad driving cycle for passenger car to model vehicle exhaust emissions calibrated using on-board measurements. Sustain. Cities Soc. **36**, 12–20 (2018). https://doi.org/10.1016/j.scs.2017.09.034

34. Yu, L., Wang, Z., Qiao, F., Qi, Y.: Approach to development and evaluation of driving cycles for classified roads based on vehicle emission characteristics. Transp. Res. Record: J. Transp. Res. Board **2058**, 58–67 (2008). https://doi.org/10.3141/2058-08

35. Quirama, L.F., Giraldo, M., Huertas, J.I., Jaller, M.: Driving cycles that reproduce driving patterns, energy consumptions and Tailpipe emissions. Transp. Res. D Transp. Environ. **82**, 102294 (2020). https://doi.org/10.1016/j.trd.2020.102294

36. Nguyen, Y.-L.T., Nghiem, T.-D., Le, A.-T., Bui, N.-D.: Development of the typical driving cycle for buses in Hanoi, Vietnam. J. Air Waste Manag. Assoc. **69**, 423–437 (2019). https://doi.org/10.1080/10962247.2018.1543736

37. Hongwen, H., Jinquan, G., Jiankun, P., Huachun, T., Chao, S.: Real-time global driving cycle construction and the application to economy driving pro system in plug-in hybrid electric vehicles. Energy **152**, 95–107 (2018). https://doi.org/10.1016/j.energy.2018.03.061

38. Yu, S., Lü, L.: Research on the influence factors of real driving cycle with statistical analysis and dynamic time warping. IET Intel. Transp. Syst. **13**, 286–292 (2019). https://doi.org/10.1049/iet-its.2018.5275

39. Hung, W.T., Tong, H.Y., Lee, C.P., Ha, K., Pao, L.Y.: Development of a practical driving cycle construction methodology: a case study in Hong Kong. Transp. Res. D Transp. Environ. **12**, 115–128 (2007). https://doi.org/10.1016/j.trd.2007.01.002

40. Jing, Z., Wang, G., Zhang, S., Qiu, C.: Building Tianjin driving cycle based on linear discriminant analysis. Transp. Res. D Transp. Environ. **53**, 78–87 (2017). https://doi.org/10.1016/j.trd.2017.04.005

41. Tong, H.Y., Tung, H.D., Hung, W.T., Nguyen, H.V.: Development of driving cycles for motorcycles and light-duty vehicles in Vietnam. Atmos. Environ. **45**, 5191–5199 (2011). https://doi.org/10.1016/j.atmosenv.2011.06.023

42. Lin, C., Zhao, L., Cheng, X., Wang, W.: A DCT-based driving cycle generation method and its application for electric vehicles. Math. Probl. Eng. **2015**, 1–13 (2015). https://doi.org/10.1155/2015/178902

43. Nyberg, P., Frisk, E., Nielsen, L.: Driving cycle equivalence and transformation. IEEE Trans. Veh. Technol. **66**, 1963–1974 (2017). https://doi.org/10.1109/TVT.2016.2582079

44. Sun, Y., Xu, H., Wu, J., Hajj, E.Y., Geng, X.: Data processing framework for development of driving cycles with data from SHRP 2 naturalistic driving study. Transp. Res. Record: J. Transp. Res. Board **2645**, 50–56 (2017). https://doi.org/10.3141/2645-06

45. Gh, M.M., Naghizadeh, M.: Development of the Tehran car driving cycle. Int. J. Environ. Pollut. **30**, 106 (2007). https://doi.org/10.1504/IJEP.2007.014506

46. Zhao, J., Gao, Y., Guo, J., Chu, L.: The creation of a representative driving cycle based on intelligent transportation system (ITS) and a mathematically statistical algorithm: a case study of Changchun (China). Sustain. Cities Soc. **42**, 301–313 (2018). https://doi.org/10.1016/j.scs.2018.05.031

47. Zhang, B., Zhang, X., Xi, L., Sun, C.: Development of a representative operation cycle characterized by dual time series for Bulldozers. Proc. Inst. Mech. Eng. Part D: J. Autom. Eng. **231**, 1818–1828 (2017). https://doi.org/10.1177/0954407016687452

48. Huertas, J., Quirama, L., Giraldo, M., Díaz, J.: Comparison of three methods for constructing real driving cycles. Energies (Basel) **12**, 665 (2019). https://doi.org/10.3390/en12040665

49. Zhang, J., Wang, Z., Liu, P., Zhang, Z., Li, X., Qu, C.: Driving cycles construction for electric vehicles considering road environment: a case study in Beijing. Appl. Energy **253**, 113514 (2019). https://doi.org/10.1016/j.apenergy.2019.113514

50. Amirjamshidi, G., Roorda, M.J.: Development of simulated driving cycles for light, medium, and heavy duty trucks: case of the Toronto Waterfront area. Transp. Res. D Transp. Environ. **34**, 255–266 (2015). https://doi.org/10.1016/j.trd.2014.11.010

51. Huertas, J.I., Díaz, J., Cordero, D., Cedillo, K.: A new methodology to determine typical driving cycles for the design of vehicles power trains. Int. J. Interact. Des. Manuf. (IJIDeM) **12**, 319–326 (2018). https://doi.org/10.1007/s12008-017-0379-y

52. Esser, A., Zeller, M., Foulard, S., Rinderknecht, S.: Stochastic synthesis of representative and multidimensional driving cycles. SAE Int. J. Alternat. Powertrains **7**, (2018). https://doi.org/10.4271/2018-01-0095

53. Hou, D., Sun, Q., Bao, C., Cheng, X., Guo, H., Zhao, Y.: An all-in-one design method for plug-in hybrid electric buses considering uncertain factor of driving cycles. Appl. Energy **253**, 113499 (2019). https://doi.org/10.1016/j.apenergy.2019.113499

54. Maamria, D., Gillet, K., Colin, G., Chamaillard, Y., Nouillant, C.: Optimal predictive eco-driving cycles for conventional, electric, and hybrid electric cars. IEEE Trans. Veh. Technol. **68**, 6320–6330 (2019). https://doi.org/10.1109/TVT.2019.2914256

55. Shen, P., Zhao, Z., Li, J., Zhan, X.: Development of a typical driving cycle for an intra-city hybrid electric bus with a fixed route. Transp. Res. D Transp. Environ. **59**, 346–360 (2018). https://doi.org/10.1016/j.trd.2018.01.032

56. Yang, Y., Li, T., Hu, H., Zhang, T., Cai, X., Chen, S., Qiao, F.: Development and emissions performance analysis of local driving cycle for small-sized passenger cars in Nanjing, China. Atmos. Pollut. Res. **10**, 1514–1523 (2019). https://doi.org/10.1016/j.apr.2019.04.009

57. Chauhan, B.P., Joshi, G.J., Purnima, P.: Candidate driving cycle construction for emission estimation. In: Mathew, T.V., Joshi, G.J., Velaga, N.R., Arkatkar, S. (eds.) Proceedings of the Transportation Research. Springer Singapore, Singapore, pp. 85–97 (2020)

58. Barlow, T., Latham, S., McCrae, I.S., Boulter, P.: A reference book of driving cycles for use in the measurement of road vehicle emissions. TRL Published Project Report (2009)

59. André, M., Joumard, R., Vidon, R., Tassel, P., Perret, P.: Real-world European driving cycles, for measuring pollutant emissions from high- and low-powered cars. Atmos. Environ. **40**, 5944–5953 (2006). https://doi.org/10.1016/j.atmosenv.2005.12.057

60. Brady, J., O'Mahony, M.: Development of a driving cycle to evaluate the energy economy of electric vehicles in urban areas. Appl. Energy **177**, 165–178 (2016). https://doi.org/10.1016/j.apenergy.2016.05.094

61. Zhang, M., Shi, S., Lin, N., Yue, B.: High-efficiency driving cycle generation using a Markov chain evolution algorithm. IEEE Trans. Veh. Technol. **68**, 1288–1301 (2019). https://doi.org/10.1109/TVT.2018.2887063

62. Arun, N.H., Mahesh, S., Ramadurai, G., Shiva Nagendra, S.M.: Development of driving cycles for passenger cars and motorcycles in Chennai, India. Sustain. Cities Soc. **32**, 508–512 (2017). https://doi.org/10.1016/j.scs.2017.05.001

63. Galgamuwa, U., Perera, L., Bandara, S.: Developing a general methodology for driving cycle construction: comparison of various established driving cycles in the world to propose a general approach. J. Transp. Technol. **05**, 191–203 (2015). https://doi.org/10.4236/jtts.2015.54018

64. Huertas, J.I., Quirama, L.F., Giraldo, M.D., Diaz, J.: Comparison of driving cycles obtained by the micro-trips, Markov-chains and MWD-CP methods. Int. J. Sustain. Energy Plan. Manag. **22** (2019). https://doi.org/10.5278/ijsepm.2554

65. Tietge, U., Mock, P., Franco, V., Zacharof, N.: From laboratory to road: modeling the divergence between official and real-world fuel consumption and CO_2 emission values in the German passenger car market for the years 2001–2014. Energy Policy **103**, 212–222 (2017). https://doi.org/10.1016/j.enpol.2017.01.021

66. Seedam, A., Satiennam, T., Radpukdee, T., Satiennam, W.: Development of an onboard system to measure the on-road driving pattern for developing motorcycle driving cycle in Khon Kaen City, Thailand. IATSS Res. **39**, 79–85 (2015). https://doi.org/10.1016/j.iatssr.2015.05.003

67. He, H., Guo, J., Zhou, N., Sun, C., Peng, J.: Freeway driving cycle construction based on real-time traffic information and global optimal energy management for plug-in hybrid electric vehicles. Energies (Basel) **10**, 1796 (2017). https://doi.org/10.3390/en10111796

68. Hu, J., Frey, H.C., Sandhu, G.S., Graver, B.M., Bishop, G.A., Schuchmann, B.G., Ray, J.D.: Method for modeling driving cycles, fuel use, and emissions for over snow vehicles. Environ. Sci. Technol. **48**, 8258–8265 (2014). https://doi.org/10.1021/es501164j

69. Han, D.S., Choi, N.W., Cho, S.L., Yang, J.S., Kim, K.S., Yoo, W.S., Jeon, C.H.: Characterization of driving patterns and development of a driving cycle in a military area. Transp. Res. D Transp. Environ. **17**, 519–524 (2012). https://doi.org/10.1016/j.trd.2012.06.004

70. Tong, H.Y., Hung, W.T.: A framework for developing driving cycles with on-road driving data. Transp. Rev. **30**, 589–615 (2010). https://doi.org/10.1080/01441640903286134

71. Triantafyllopoulos, G., Dimaratos, A., Ntziachristos, L., Bernard, Y., Dornoff, J., Samaras, Z.: A study on the CO_2 and NOx emissions performance of Euro 6 diesel vehicles under various chassis dynamometer and on-road conditions including latest regulatory provisions. Sci. Total. Environ. **666**, 337–346 (2019). https://doi.org/10.1016/j.scitotenv.2019.02.144

72. Yang, Z., Deng, B., Deng, M., Huang, S.: An overview of chassis dynamometer in the testing of vehicle emission. MATEC Web Conf. **175** (2018)

73. Li, T., Chen, X., Yan, Z.: Comparison of fine particles emissions of light-duty gasoline vehicles from chassis dynamometer tests and on-road measurements. Atmos. Environ. **68**, 82–91 (2013). https://doi.org/10.1016/j.atmosenv.2012.11.031

74. Kousoulidou, M., Fontaras, G., Ntziachristos, L., Bonnel, P., Samaras, Z., Dilara, P.: Use of portable emissions measurement system (PEMS) for the development and validation of passenger car emission factors. Atmos. Environ. **64**, 329–338 (2013). https://doi.org/10.1016/J.ATMOSENV.2012.09.062

75. Franco, V., Kousoulidou, M., Muntean, M., Ntziachristos, L., Hausberger, S., Dilara, P.: Road vehicle emission factors development: a review. Atmos. Environ. **70**, 84–97 (2013). https://doi.org/10.1016/j.atmosenv.2013.01.006

76. Chen, L., Wang, Z., Liu, S., Qu, L.: Using a chassis dynamometer to determine the influencing factors for the emissions of Euro VI vehicles. Transp. Res. D Transp. Environ. **65**, 564–573 (2018). https://doi.org/10.1016/j.trd.2018.09.022

Design and Implementation of Monitoring Campaigns to Obtain Driving Patterns

4

José I. Huertas, Jenny Díaz Ramírez, Luisa F. Chaparro Sierra, Oscar S. Serrano-Guevara, and Daniel Cordero-Moreno

Abstract

This chapter aims to outline a comprehensive methodology for designing and implementing monitoring campaigns focused on capturing the driving patterns of a specific region or application, covering common approaches used at each stage of the process. The chapter emphasizes the first two stages—design and implementation—of the overall methodology for constructing driving cycles (DCs). Special attention is given to vehicle selection, sample size determination, and vehicle instrumentation prior to conducting the monitoring campaign. Additionally, we summarize preprocessing and analysis techniques employed to manage the collected data.

J. I. Huertas (✉) · J. Díaz Ramírez · L. F. Chaparro Sierra · O. S. Serrano-Guevara
Sustainable Energy Research Group, Tecnologico de Monterrey, Monterrey 64849, México
e-mail: jhuertas@tec.mx

J. Díaz Ramírez
e-mail: jdiaz@tec.mx

L. F. Chaparro Sierra
e-mail: lchaparr@tec.mx

O. S. Serrano-Guevara
e-mail: oserrano@tec.mx

D. Cordero-Moreno
Universidad del Azuay, Cuenca, Ecuador
e-mail: dacorderom@uazuay.edu.ec

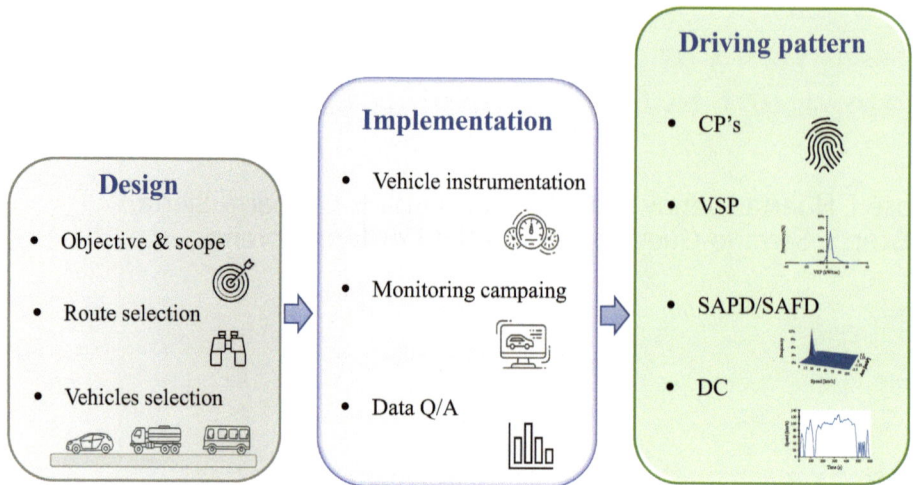

Fig. 4.1 General methodology for obtaining driving patterns

4.1 Methodology for Obtaining Driving Patterns

Figure 4.1 illustrates the stages of the methodology for obtaining driving patterns, which align with most of the reviewed studies. We highlight that many studies on driving patterns or driving cycles do not encompass all the steps outlined in Fig. 4.1. Instead, they tend to focus on one or two stages of the overall methodology, often overlooking or underestimating the importance of the others. In particular, steps related to defining the monitoring campaign's objectives and ensuring representativeness are frequently incomplete. Chapter 5 will delve into the description of driving patterns.

4.2 Defining Objectives and Scope of the Monitoring Campaign

As a first step, it is essential to identify the motivating factors behind identifying driving patterns and the intended applications for their use.

Often, the interest in driving patterns arises from governmental authorities seeking to develop a DC that accurately represents the local driving pattern as part of the vehicle homologation process. In this context, the primary objective of the monitoring campaign is to create a DC that tests the energy and environmental performance of vehicles under local conditions, ensuring they meet predefined standards before being approved for sale in the local market.

Alternatively, the objective might be to identify the typical loads experienced by vehicles in a specific region or application, enabling optimization of the powertrain or adaptation to local conditions. Another goal of a monitoring campaign could be to assess driving styles within a transport service company as part of an eco-driving program, with the aim of identifying strategies to reduce vehicle energy consumption.

4.3   Study Region

The second step in designing a monitoring campaign to capture driving patterns involves a thorough description and understanding of the study region. Particular attention should be paid to factors that influence driving patterns, such as location, topography, weather conditions, road conditions, and vehicle fleet composition. Socio-cultural factors, including income per capita, average educational level, and the age of the population, are also important considerations. For example, analyzing the driving patterns of professional drivers in public transportation differs significantly from studying those of informal taxi drivers. Understanding these regional characteristics becomes relevant when applying the obtained driving pattern for practical actions oriented toward the fulfillment of the objectives set for the monitoring campaign.

Despite the relevance of describing the region of interest, nearly half of the references reviewed do not mention the region under study in their abstracts. Moreover, at least 13% of the documents in the corpus omit any mention of the study region entirely. This omission may be acceptable when the primary purpose of the study is not to deriving a specific driving pattern but to demonstrating a methodological aspect of the DC construction process.

Most of the regions studied are urban centers, with some studies extending to highways that connect multiple cities, and only one study encompassing more than one country. Figure 4.2 illustrates the number of studies published per country, highlighting a significant interest of Asian countries in the development of driving patterns, followed by North America and Europe. Table 4.1 details the attributes included in the description of each study region.

4.4   Routes Selection

The selection of the routes must align with the monitoring campaign's objectives. A route comprises the collection of road segments that vehicles follow while their operational conditions are monitored. It is widely recognized that route selection is one of the most critical and subjective steps when obtaining driving patterns [21, 44]. Figure 4.3 shows several approaches observed in the literature to select routes with the purpose of grasping

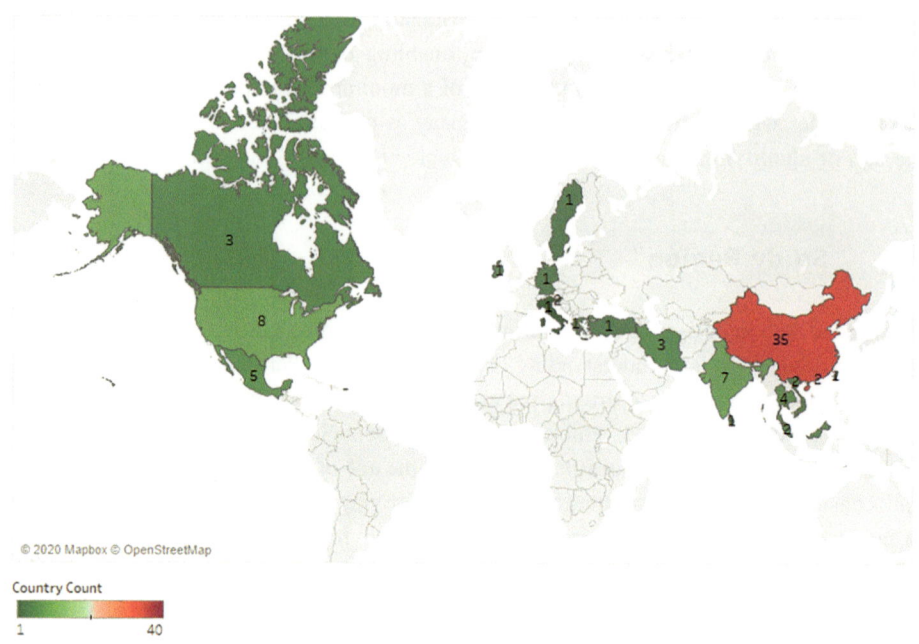

Fig. 4.2 Number of publications per country where monitoring campaigns to obtain driving patterns have been carried out by 2020

driving patterns: predetermined routes, main routes, a mix of road types, and undefined routes. Next, we will describe each of these approaches.

Predefined Routes

Most of the time, the routes where the monitoring campaign will be conducted are predefined as part of the study's scope. For example, the study's scope may encompass the fixed routes where public buses operate within the city [28, 36, 40]. In other instances, the study's objective is methodological, utilizing information from a single route to validate the method rather than offering a driving pattern for the region [45]. In some instances, researchers' expertise and knowledge are used to select routes within the study area. At least twenty references use a single route or corridor (e.g., [10, 20, 30, 35, 41, 46–48]), while approximately ten other studies select from 2 to 21 routes to represent the study region (e.g., [10, 20, 30, 41, 47, 48]). However, this arbitrary route selection approach can potentially introduce bias into the collected data [49].

Routes with the Maximum Traffic Flow

The second approach to selecting the routes for the monitoring campaigns uses traffic flow as a criterion. The routes with the highest traffic flow in the region of interest are

Table 4.1 Common descriptive attributes that are included in the description of the study region

Attribute	Description	Example references from set C1
Region	City or zone under study	Almost half of the corpus
Population	Amount, gender, etc.	[1–3]
Roads	Specific routes where the monitoring campaign takes place. (e.g., Highway, route, circuits, or roads)	[1, 4–13]
Length /size	Monitoring campaign time, sample size, or distance.	[5, 14–16] [7, 8, 16, 17, 17, 18, 18, 19, 19–25]
Fleet description	Type of vehicles used in the collection phase (e.g., buses, light vehicles, motorcycles, etc.)	[1, 2, 6, 10, 20, 23, 26]
Traffic	Traffic volume	[4, 5, 11, 21, 26–28]
Road type	Level of service (LoS), road classification, speed limits	[20–22, 29–33]
Road description	Number of traffic lights, stops, or stations, road grade or altitude, etc.	[32, 34, 35] [23, 26, 29, 36–38]
Travel activity	Origin-destination patterns, land use, driving behavior, speed, congestion	[9, 17, 39] [2, 8, 10, 29, 40]
Times	Specific periods in the day to distinguish different levels of activity (e.g., peak, off-peak, schedule)	[5, 18, 22, 26, 37, 41, 42] [26, 27, 37, 41–43]
Localization software	e.g., GoogleMaps, ArcGis, GeoJson, OpenStreetMap	[5, 22, 26, 41, 42]

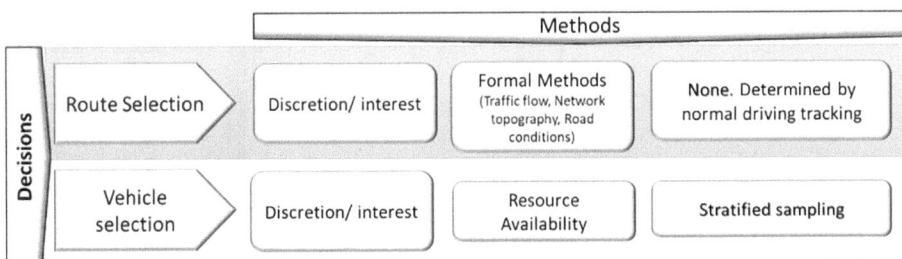

Fig. 4.3 Illustration of the approaches used to select the routes that the vehicles will follow while their operational conditions are monitored with the purpose of grasping driving patterns

selected. It uses formal traffic flow models [9, 21, 39, 46], official reports of traffic conditions (speed profiles, congestion, etc.), traffic flow surveys (origin-destination travel patterns, travel times, etc.), and descriptions of the topographic structure of the road network (levels of service of roads, road types, topography, land use, etc.). The selection process must ensure that the set of routes or paths selected shows a traffic behavior representative of the region in terms of length, proportion of road times, speed, traffic, and road types, among other traffic indicators.

Routes that Sample Diverse Types of Roads

This approach involves designing routes that sample the various types of roads available within the region of interest. However, there is no single, universally accepted way of classifying roads. Table 4.2 provides examples of the diverse options for classifying roads. The first two rows of the table are classifications used in some of the references, and they carefully consider road types when selecting routes. The last row describes some national standards. A review of road classifications by country can be found in [50]. National systems of road classifications consider eight characteristics to classify roads, which are grouped into four criteria: form (access control and road surface), usage, designation (administration, link role, place status, transport mode), and performance (function) [50]. European and North American road classifications are among the most complete ones, covering at least one characteristic per criterion.

We recommend the use of the US level of service classification. Level of service (LoS) is a qualitative measure of quality service that the road offer to the transit of motor

Table 4.2 Systems of road classification

Per areas [26] • Urban • Rural	Based on their shape [10] • Circular • Radial • Straight	Based on LoS [34] • Urban • Uphill • Mountain	Based on road conditions [47] • Paved • Unpaved
[33] • Urban • Suburban • Viaduct • Freeway	[21] • Expressway • Main road • Secondary road • Branch road	[30] • Urban • Suburb • Highway	Based on traffic density [29] • High • Low
Europe [51] • All-purpose express I and II • Motorways • Express roads	UK [52] • Motorway • Primary A-road • B-road • Classified and unclassified • Unofficial: C-road, major and minor roads	China [53] • Expressway • Classes I–IV	US [54] Based on the volume-to-capacity ratio: • Classes A to F

vehicles. LoS is used to analyze roadways and intersections by categorizing traffic flow and assigning quality levels of traffic based on performance measures like vehicle speed, density, congestion, etc. Six categories are considered:

- A: free flow. Traffic flows at or above the posted speed limit, and motorists have complete mobility between lanes.
- B: reasonably free flow. LoS at speed limits are maintained, and maneuverability within the traffic stream is slightly restricted.
- C: stable flow, at or near free flow. The ability to maneuver through lanes is noticeably restricted, and lane changes require more driver awareness.
- D: approaching unstable flow. Speeds slightly decrease as traffic volume slightly increases. Freedom to maneuver within the traffic stream is much more limited, and driver comfort levels decrease.
- E: unstable flow, operating at capacity. Flow becomes irregular, speed varies rapidly due to the absence of usable gaps to maneuver in the traffic stream, and speeds rarely reach the posted limit.
- F: forced or breakdown flow. Every vehicle moves in lockstep with the vehicle in front of it, with frequent slowing required. Travel time cannot be predicted, with generally more demand than capacity. A road in a constant traffic jam is at this LoS.

However, the LoS ignores the surface aspects of the road and road grades, which are relevant in Latin American (LATAM) countries. It does not provide a quantitative method to grade roads.

Normal Users' Routes
Finally, the fourth approach, which is our recommendation, consists of not selecting a route beforehand, allowing the vehicle to move freely around the road network without any restriction. This approach has been considered the most suitable when developing driving cycles for large regions with complex road networks [49]. However, this approach requires monitoring a large sample of vehicles for a long period. It has become the most preferred alternative, especially after the massification of telemetry systems to track the operation of vehicles [1, 2, 5, 9, 14, 16, 55–59].

4.5 Vehicle Selection

When dealing with driving patterns, it is important to specify the type of vehicles targeted. Depending on the purpose of the study, the decision regarding vehicle selection varies and is typically influenced by the researcher's discretion or resource availability. It goes from following the behavior of a single vehicle [57, 60, 61], selecting a subset of the fleet [9, 14], to a large set of volunteered vehicles [55].

In LATAM, light-duty vehicles (LDV) are mostly gasoline-fueled and correspond to passenger cars. Some countries add bioethanol (5, 10, or 20%) to gasoline. Meanwhile, heavy-duty (HDV) vehicles are diesel-fueled and used for the collective transport of passengers and freight transport. HDVs are vehicles with a gross weight greater than 2.4 tons. Small percentages of vehicles use natural gas (CNG), liquified petroleum gas (LPG), or electricity as their source of energy.

The absence of a unique or standard vehicle classification arises from the need of each local governmental authority to establish its own vehicle classification for regulatory purposes. Thus, names assigned to vehicle types could vary from country to country. Most vehicles are classified depending on their use (passenger, buses, and trucks) and the number of axles. Table 4.3 shows the classification in Mexico for heavy-duty vehicles (HDV). Table 4.4 shows the vehicle type selected in the corpus of the works consulted in the literature review.

Table 4.3 Classification of heavy-duty vehicles in Mexico [62]

Transport	Name	Number of axles	Number of wheels	Vehicle configuration	Gross vehicle weight rate [t]
Passengers	Van	2	4		6.5
	B2	2	6		19
	B3	3	8–10		24–27.5
	B4	4	10		30.5
Cargo	C2	2	It depends on the configuration of the motor unit		19
	C3	3			24–27.5
	T2	3			30–45.5
	T3	3			46.5–54.5

Table 4.4 Summary of vehicles selected in works that deal with driving patterns and cycles.

Type of vehicle		Example references
Buses		[4, 9, 34, 38, 41, 63]
Electric vehicles		[4, 16, 21, 29, 64–67]
Light vehicles		[11, 33, 42, 55, 60, 61, 68, 69]
Others	– Military	[47]
	– E-motorcycles	[70]
	– Motorcycles	[8, 17, 26]
	– Other specific activities	[45, 71]

4.6 Sample Size

This step aims to define the sample size that needs to be monitored to ensure that the obtained driving pattern represents the study's region or application. Sample sizes are usually reported in the works consulted in the literature but are not necessarily justified or addressed appropriately. In all cases, the discussion of sample size was limited to the number of vehicles used in the monitoring campaign. We will show that this question remains open for additional research work. Next, we describe the partial answers found so far.

Given that the topic of representativity often leads to frequent confusion when applied to vehicle-related assessments (e.g., emission factors, specific fuel consumption, etc.), we start by providing relevant definitions.

Variables of interest are the quantitative variables that need to be evaluated with the sample of vehicles. For example, fuel consumption (L/km) and tailpipe emissions (g/km). Driving patterns cannot be used as a variable of interest because they are not a quantitative variable. Instead, the characteristic parameters that describe driving patterns can be used as variables of interest, such as vehicle speed, positive acceleration, and percentage of idling time.

Population: All vehicles in the region of application.

Sample: A subset of vehicles from the population.

Representative sample: A sample possessing the same characteristics as the population. The characteristics of the population that refer to this definition are those closely related to the variable of interest. For example, suppose that the variable of interest is fuel consumption. In that case, the sample should include vehicles with diverse technological characteristics that influence fuel consumption in proportions similar to the observed in the population. Note that this definition of a representative sample is qualitative and does not impose requirements on the sample size. This definition does not allow the quantification of the degree of representativity of a sample.

Sampling Method: The method followed in selecting the individual vehicles that belong to the sample ensures the reproduction of the fleet composition in terms of the variables influencing the variable of interest. It can be probabilistic or non-probabilistic. When using probabilistic methods or random sampling, the sample is expected to be representative of the population, independent of its size, because the characteristics of the population would be represented in the selected sample. Therefore, the conclusions can be extrapolated to the entire population. Here, random refers to the way to select elements (in this case, vehicles) from the population, in which each one has the same (or a known) probability of being selected (simple vs stratified random sampling). However, this method is more time-consuming and expensive. Systematic and clustering methods can also be applied to choose the elements of the sample. Non-probabilistic sampling methods should be used with the precaution that the conclusions will be limited to the sample chosen and to the vehicles that match the sample characteristics selected during the study.

Accuracy of the results: The dispersion or variability of the results concerning the estimated mean of the variable of interest ($\bar{x}$). It can be described as:

- The confidence interval of the estimated mean ($CI\mu$) with a confidence level ($1 - \alpha$) (Eq. 4.1).
- The uncertainty (I, Eq. 4.2) corresponds to the mean range of the confidence interval. Mean estimations are equivalent to the standard error of the mean (SE) multiplied by the critical value of the normal distribution or t-distribution.
- The percentage error (e) is the ratio of uncertainty I to the estimated mean value (Eq. 4.3).

$$IC\mu = \bar{x} \pm I \tag{4.1}$$

$$I = t_{a/2,n-1}\frac{s}{\sqrt{n}} \tag{4.2}$$

$$e = \frac{I}{\bar{x}} \tag{4.3}$$

The *degree of statistical representativity* is measured through percentage error (e) with a given confidence level. Values close to zero correspond to higher levels of representativity. It is emphasized that the true average value of the population, with zero uncertainty or 100% representativity, can only be obtained by sampling 100% of the population.

Representative sample size. Equation 4.4, which was derived from Eq. 4.3, computes the sample size required to assure representativeness given an acceptable percentage error with a given confidence level.

$$n = \frac{Z_\alpha^2 N p q}{e^2 (N - 1) + Z_\alpha^2 p q}$$
(4.4)

where:

n	is the sample size,
N	is the population size,
Z_α^2	the critical value for the confidence level $(1 - \alpha)$,
p	is the proportion of individuals in the population with the study characteristic,
q	is the proportion of individuals without that characteristic (i.e., $q = 1 - p$).

These data are generally unknown, and it is often assumed that $(p = q = 0.5)$. This equation is used when the standard deviation is unknown.

Sample size for the case of variables of interest that depend only on vehicle technology
Note that Eqs. 4.1–4.4 apply only when the parameter of interest (e.g., the mean) is assumend to follow a normal distribution. Researchers often use Eq. 4.4 to calculate the number of vehicles required to ensure that the experimentally determined value of the variable of interest obtained with the reduced sample represents the value of the variable of interest of the whole population. According to Eqs. 4.1–4.3, the larger the sample size, the smaller the error percentage (e). Furthermore, Eq. 4.4 becomes insensitive to population sizes (N) greater than ~7,500. Additionally, for population sizes greater than this value, with a 95% confidence level (a) and with an uncertainty or relative error of $e < 10\%$, Eq. 4.4 estimates that 68 vehicles are required to assure representativeness per category of vehicles where the parameter of interest exhibit a normal distribution.

However, in the case of vehicles, the use of Eq. 4.4 has several important implications. First, the assumption of a single distribution of the variable of interest is uncommon in the case of vehicles. Figure 4.4 shows, as an example, the frequency distributions of fuel consumption of 54 t trucks used for freight transport under normal conditions of use in Mexico. It shows that the fuel consumption of the whole population, when expressed in terms of L/km, does not exhibit a normal distribution. Not all plots fall into a single distribution.

The lack of uniformity of the results (results distributed around a single mean value) is because the variable of interest depends on several influencing factors. For example, vehicle fuel consumption varies with vehicle weight (Fig. 4.4c) and vehicle year model (Fig. 4.4b). In this case, the entire population should be segmented into categories according to the influencing variables. Therefore, in opposition to applying Eq. 4.4 to the whole population of vehicles where the variable of interest does not satisfy the requirement of normal distribution, it should be used independently for each vehicle category where the variable of interest does exhibit a normal distribution. For example, instead of applying Eq. 4.4 to the whole population of 54 t HDV in Fig. 4.4a, it could be used for the

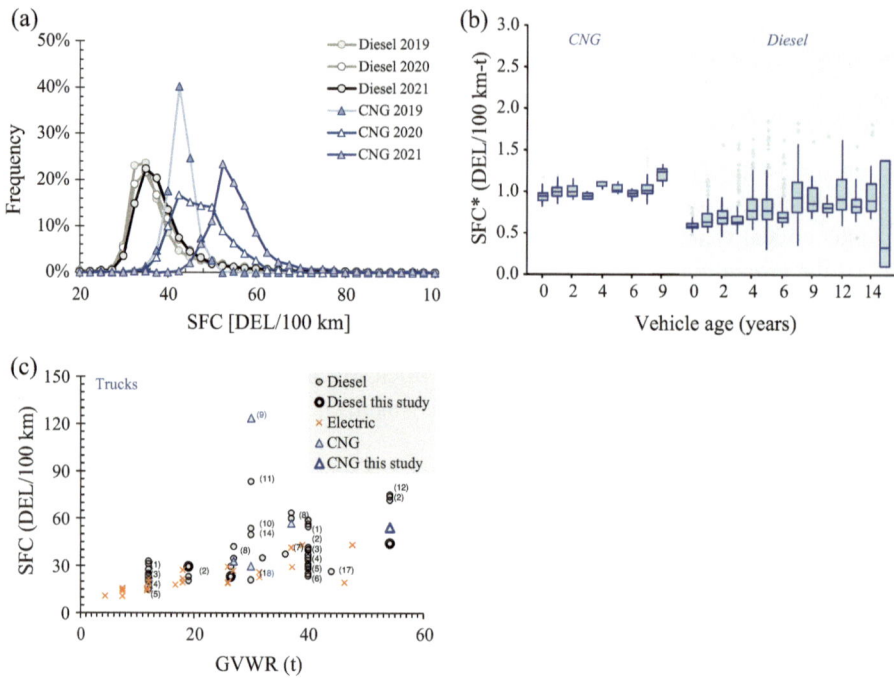

Fig. 4.4 Illustration of alternatives used to report results of vehicles' fuel consumption. **a** Frequency distribution of the specific fuel consumption (SFC, expressed in L/100 km) observed in 54 t, diesel, and CNG HDV trucks used for long-range freight transport under normal conditions of use in Mexico. **b** The same results, expressed in terms of L/100 t km, box plots as a function of vehicle age. **c** Average specific fuel consumption of vehicles as a function of gross vehicle weight rate (GVWR)

diesel and CNG 54 t HDV category. Furthermore, the CNG vehicle category should be subdivided according to the vehicle model year.

Alternatively, the variable of interest can be modified to eliminate the effect of the influencing factors. For example, the fuel consumption in terms of L/km can be divided by the total vehicle weight. The new variable expressed in (L/t km) groups several categories of vehicles.

Size of the Sample for the Case of Grasping Driving Patterns

There is no easy alternative to quantify the minimum size of the sample needed to grasp a driving pattern in a given region.

First, a driving pattern is not a quantitative variable and, therefore, needs to be described through a combination of characteristic parameters such as speed acceleration, percentage of idling time, etc. Thus, in this case, the variables of interest are those characteristic parameters. Second, driving patterns are influenced by vehicle technology, driver habits, and external factors (route and weather conditions). It means that a sample combines a vehicle, a driver, and a route under specific external conditions. Thus, the question is how many combinations of these four variables are needed to represent the driving pattern of the existing combinations of these variables in a given region. There is no need to keep trying to elucidate an answer to this question because soon, the conclusion is that any realistic sample size, with today's technology, will have a low (and unknown) level of representativeness.

Researchers decide on the sample size according to economic and time resource limitations. They include as many vehicles and drivers as possible and extend the monitoring campaign as long as possible to include the many external conditions in the region under study.

4.7   Vehicle Instrumentation

In this step, the vehicles selected in the previous step are instrumented. Monitoring the vehicle location, speed, and altitude at a frequency of 1 Hz is relevant to capturing the driving pattern in a given region or application. Nowadays, recording vehicles' instant fuel or energy consumption is highly desirable. Shortly, additional variables will be required, especially the load transported and the tailpipe emissions. Table 4.5 shows the instruments most frequently used to carry out these tasks.

Speed and Location
Global position systems (GPS) are currently the most frequently used instruments to monitor the location and speed of vehicles. GPS were released for civil use in 2000. Later, differential GPS, which improved GPS's location accuracy from meters to a few centimeters, achieved the appropriate spatial resolution required for practical applications, including the description of driving patterns. GPS steadily decreased in cost, ranging from a few thousand to a few USD. Moreover, these systems have the advantage of being non-invasive and discreetly located not to affect the driver's everyday habits [28]. Various GPS technology-based devices, such as GSM (Global System for Mobile Communication), digital road map-based navigation systems, and AVL (advanced vehicle location), can also be implemented to ease data processing.

Altitude
Typically, authors interested in driving patterns conduct their research in regions with flat terrain. Currently, there is a lot of interest in regions with significant variations in altitude, such as in LATAM countries, where the world's ten highest cities are located. In these cases, it is important to monitor the altitude at which the monitored vehicle is located. Although GPS technology reports altitude, the reported data is highly inaccurate

Table 4.5 Variables and instruments used to monitor the instant operative variables of vehicles with the purpose of capturing driving patterns.

Variable	Options	Advantages	Disadvantages
Location and Speed	Speed sensors (fifth-wheel)	High accuracy	High cost The location is not recorded
	GPS	Commercial availability Low cost	Loss of signal in tunnels and bridges High cost if high resolution is required
	OBD	Measurement accuracy Included in telemetry systems	Restricted to some vehicle technologies
Altitude	GPS*	Commercial availability Low cost Low accuracy	Loss of signal in tunnels and bridges High cost if high resolution is required
	Digital elevation model	Based on [72]	Free
Fuel consumption	OBD	Instantaneous consumption	Costs
	Fuel flowmeters	Low costs	Low accuracy
Tailpipe emissions	PEMS		
Load transported	Strain gauges	N/A	No available commercially for massive use in vehicles

due to technical limitations in the altitude determination method. This information is particularly relevant because altitude variations demand power from vehicles, increasing fuel consumption and vehicle emissions. For well-established routes, altitude can be determined using theodolites. However, this process is highly labor-intensive. Alternatively, we suggest consulting official cartographic information, which is usually available to the public in each country. For each latitude and longitude position on a route, these sources of information provide altitude with great accuracy. We recommend obtaining the instant vehicle altitude using the GPS Visualizer Digital Elevation Model (DEM) (www.gpsvisualizer.com/elevation). Additionally, we recommend using tools such as Regulation (EU) 106/646 to improve altitude quality and slope definition. This methodology was validated in [73], where authors compared results from this tool with official values reported by government institutions obtained by on-site measurements.

Fuel Consumption

The fuel consumption in vehicles can be monitored manually by recording the whole trip's fuel consumption. Transport companies and/or vehicle owners have developed platforms where fleet operators report the amount of fuel loaded and the distance traveled until the next fill-up. Results from this alternative can be biased by third parties involved in the fuel supply business. It can also be obtained by installing inline flow meters.

The measurement of fuel consumption in vehicles used to be a cumbersome process until the advent of the fuel injection system. In the 1990s, electronic fuel injection systems replaced carburetors to reduce vehicle pollutant emissions. This system relies on an Engine Control Unit (ECU), which is a computer that reads the various sensors installed in the engine to determine the amount of fuel injected into the combustion chamber. Among the engine sensors are those for engine speed (RPM) and load measured through the Manifold Absolute Pressure (MAP) sensor. The ECU regulates fuel injection by controlling the injection time.

To simplify engine maintenance, manufacturers introduced a port where the information gathered by engine sensors and collected by the ECU could be shared via the Controlled Area Network (CAN) protocol with external devices for diagnostic purposes. This system was named the On-Board Diagnostic System (OBD) and was standardized as OBD II, becoming mandatory in the United States in 1996 and Europe in 2000. However, it wasn't until 2010 that OBD II became available to the public, with the introduction of low-cost commercial devices, for instance, ELM327, priced under USD 200, enabling OBD II readings outside service centers.

The ECU reports engine fuel consumption using any of the following methods. A direct measurement of the fuel injection time, which is directly correlated to the fuel injected into the engine combustion chamber. Alternatively, it can be estimated indirectly through the readings of the sensors that manufacturers install on the engine to control its operation. Equations 4.5 and 4.6 are some of the alternatives [73]. In these equations, $\dot{v}_f$ is the volumetric fuel consumption rate, $\dot{m}_a$ is the mass airflow, which is measured by the mass airflow (MAF) sensor, AF is the air-fuel ratio, which is measured by the lambda sensor. ρ_f is fuel density at standard conditions. P and T are the inlet air pressure and temperature which are read by the mass airflow pressure (MAP) and the intake air temperature (IAT) sensor. R_a is the air constant. V_e is the engine volume. RPM is the engine speed, and n is the number of revolutions per cycle ($n = 2$).

$$\dot{v}_f = \frac{\dot{m}_a}{AF \; \rho_f} \tag{4.5}$$

$$\dot{v}_f = \frac{1}{AF \, \rho_f} \frac{P}{R_a T} \; V_e RPM \; \frac{1}{n} \frac{2\pi}{60} \tag{4.6}$$

Using both the previously mentioned OBD II estimation method and the gravimetric reference method (fuel weight differences) simultaneously, Quirama et al. [31] replicated

Fig. 4.5 Correlation between
the fuel consumption obtained
via OBD and the standard
gravimetric method [31]

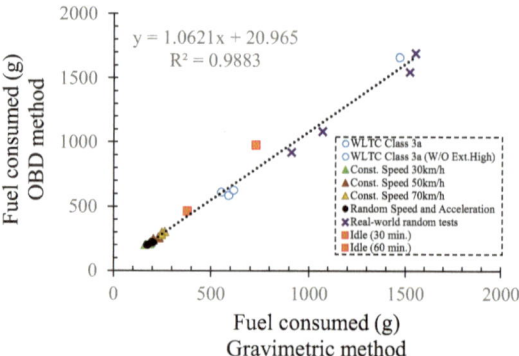

the Worldwide Harmonized Light Vehicles Test Cycle (WLTC) type-approval driving cycle on a chassis dynamometer to measure fuel consumption. They also assessed fuel consumption under random speed-acceleration tests, steady-state conditions at 30, 50, and 70 km/h in laboratory settings on a chassis dynamometer and included idle tests. Furthermore, they measured fuel consumption under real-world conditions in several 30–60 min random tests. By grouping all these tests into a correlation analysis based on the total mass of fuel consumed (Fig. 4.5), they achieved a coefficient of determination (R^2) higher than 98%, along with a slope of 1.06. This slope suggests an average difference of 6% between all the paired measurements of the conducted tests. These results strongly supported the reliability of the fuel consumption data reported by OBD.

The reading of the OBD II port has become the best alternative for monitoring fuel consumption, engine RPM, and engine load in vehicles nowadays. This method facilitates the monitoring of large samples of vehicles for extended periods.

Tailpipe Emissions

The measurement of tailpipe emissions has been challenging since its inception in the 1980s, following the establishment of the first DCs. During this period, emissions were measured with vehicles on a chassis dynamometer following a DC, and a Constant Volume Sampler (CVS) was utilized to determine the total mass of pollutants emitted during the test. This procedure is the current standard for measuring tailpipe emissions for legislative purposes.

However, criticisms of this standard protocol have emerged because it does not accurately replicate real emissions from vehicles operating under real driving conditions. Consequently, in the 2000s, researchers began measuring tailpipe emissions through road tests. The Mobile Emissions Laboratory (MEL) at the University of California in Riverside was a notable project that aimed to determine the real tailpipe emissions. It involved a CVS mounted in a trailer that effectively chased the vehicles under study. Over time, this technology evolved into a Portable Emissions Measurement System (PEMS), capable of measuring exhaust emissions data under real road conditions [22, 57, 70, 74–76].

PEMS measures the exhaust emissions in real road conditions and laboratory conditions with the vehicle running on a chassis dynamometer [22, 57, 70, 74, 75]. PEMS measures CO, NOx, HC, particulate matter (PM and PN), and NH_3. They are highly costly (0.1–1 million USD). This instrument should be calibrated before each test using NIST traceable calibration gases. Additionally, it should be calibrated after every hour of continuous operation to address potential zeroing issues with the CO and NOx sensors. These issues demand that highly trained personnel be involved in the tests with PEMS. Thus, tests with PEMS are time-consuming and highly costly.

Vehicle Weight
Currently, vehicles do not have sensors to monitor this variable, and frequently, transport companies do not keep a record of this variable. Vehicle weight strongly influences fuel consumption in freight transportation, where payload can vary widely (0–35 t). Payload is frequently monitored on the road to ensure that the circulating vehicles satisfy the local regulations regarding the maximum load transported. Currently, the local regulations in LATAM countries require companies to report the weight and type of load they transport on every trip. However, there is still a need to have access to the payload associated with each trip.

Data Acquisition Frequency
The most common data acquisition frequency is 1 Hz [1, 5, 6, 9, 28, 37, 39, 40, 48, 58, 59, 64, 71, 77, 79, 80]. Higher frequencies (2–10 Hz) are suggested to describe vehicle dynamics better, though 1 Hz is acceptable for estimating vehicle energy consumption [34]. Higher frequencies can be found in [34] at 4 Hz, [27] at 5 Hz, and [14] at 10 Hz. Typically, collected data is sent to a server either at the end of the day [14, 28] or after the data collection period [71].

4.8 Monitoring Campaign

The next step in obtaining driving patterns is collecting data through a monitoring campaign. The objective is to monitor a representative sample of the trips carried out in the region of interest using a representative sample of vehicle technologies driven by a representative sample of drivers during the entire year to grasp the season's influence. This requirement results in expensive and time-consuming monitoring campaigns.

Historically, the cost of vehicle instrumentation has driven decisions on data collection methods. Most of the time, researchers have had access to few instrumented vehicles. Then, they designed their monitoring campaigns using a combination of the following strategies.

- Scheduled routes at fixed hours: The instrumented vehicles cover fixed routes at fixed hours every day of the week.
- Chase car: The driver of the instrumented vehicle randomly picks a target vehicle and chases it.
- Professional driver: The instrumented vehicle is driven by an expert driver who covers a given route using different driving styles (aggressive, as usual, and gentle).
- Normal use: vehicles are driven by their normal user following his/her everyday routine without any intervention.

The chase car protocol was proposed by Austin et al. (1993) and was widely used until the first decade of this century [39, 81], primarily due to its low cost. In the chase car method, a target car is followed by a specially instrumented chase car driven by a trained driver. The target vehicle is randomly selected while in motion on the chosen route. Once it deviates from the route, another vehicle is chosen. The chase car is expected to maintain a constant distance from the target car during various operational modes, allowing for a time lag for both acceleration and deceleration phases [12, 48, 61, 73, 82]. Some characteristics that define this method are:

- It enables the collection of a substantial amount of data from a selected section of the route [1, 48]. However, it is limited to gathering information from random trip sections that match the chosen route [61].
- It is highly cost-efficient since only one car needs to be instrumented [21, 78].
- It presents challenges in implementation when dealing with aggressive behaviors, low or very high traffic, and the criteria for selecting the new car to be tracked [44, 48, 81].
- It may record differences in transient speed concerning the chased car [1], especially in highly sloping segments, during higher accelerations/decelerations, or when the chase car catches up with the new vehicle to be tracked [48, 81]. Nevertheless, few studies show a high correlation between the target car's speed and the chase car's speed [81]. It also has the potential to induce different driving behaviors in the chased car, which is challenging to identify from the collected data [81].

On the other hand, the normal use method involves the instrumentation of numerous vehicles and long-lasting monitoring campaigns. The main advantage of this method is that it captures the everyday driving behavior of vehicle users without external interference. The increasing availability of low-cost telemetry systems based on OBD for monitoring the operation of vehicles with limited human intervention has made this method the preferred choice in recent decades [61].

We suggest installing OBD data gathering systems in as many vehicles as possible and letting their regular driver drive them without interfering with their everyday routine. State-of-the-art information technology allows huge amounts of data gathered by the OBD systems to be transferred to the cloud at a reasonable cost in autonomous mode.

Currently, there are companies that perform this task for cargo companies. These technologies were designed to monitor vehicles' operation and warn fleet managers of eventualities in the vehicle operation, such as dangerous brake events, alerts of engine malfunctioning, and instant location of the vehicles. These technologies monitor the vehicles through OBD systems and automatically transfer information to the cloud. Large samples of vehicles (~5000) have been monitored in long-lasting monitoring campaigns (3 years) using these commercially available services.

The main drawback of the telemetry systems is that they were not originally designed for capturing driving patterns. Therefore, they record operational parameters at smaller frequencies than desired (<1 Hz). Techniques have been under development to face this issue.

Monitoring Campaign Duration
The data collected in the region under study should last long enough to include randomly all factors that influence the local driving pattern, including factors such as seasonality, trip composition, road utilization, and traffic behavior.

There is a significative dispersion in the duration of monitoring campaigns. Duration ranges from three days [61] to several months [59, 68, 83], one year [10, 16, 22, 55], two years [84] and up to three years [66]. Some studies report the duration of the monitoring campaign in terms of the amount (or time) of data collected [7, 9, 33, 45, 85]. Others report the monitoring campaign duration in terms of the number of effective trips [31, 32, 86], and others in terms of the number of micro trips or segments [20, 56, 65]. However, the decisions about the duration of the monitoring campaigns are still dominated by extrinsic factors, such as resource availability, as well as the researcher's discretion, without any robust statistical analysis regarding how long the monitoring campaigns should be to represent the region's driving pattern adequately. Thus, the estimation of errors is presented as a result, not as an experimental design decision or goal.

4.9   Data Preprocessing and Data Quality Analysis

After monitoring the instant operational conditions of a representative sample of vehicles as described in previous sections, a data preprocessing task must be accomplished to ensure data quality, integrity, and completeness. The main objective of this task is to identify erroneous data, act on incomplete data, and synchronize different data sets. This section describes the techniques most frequently used and those suggested by the authors to accomplish these objectives.

Data Discontinuity
The problem of data discontinuity or missing data often occurs to GPS, which faces issues when vehicles pass through tunnels, edges, or other physical barriers. It is also observed

when PEMS stops recording data on a trip during self-calibration [31]. It also could occur due to the loss of power to the instrumental system. A common approach to addressing missing data involves disregarding monitored trips exceeding a relative threshold (e.g., 5% or 10%) of missing data [31, 63, 86]. Alternatively, missing data can be filled with interpolated data obtained using the closest available data points [23]. Equations 4.7 and 4.8 can be used for this purpose. They are linear and quadratic interpolations, respectively. In these equations, y is the desired interpolated value at the missing position or time x in the range $x_0 < x < x_2$. And (y_0, x_0), (y_1, x_1), (y_2, x_2) are the known data points.

$$y = a_o + a_1(x - x_o) \tag{4.7}$$

$$y = a_o + a_1(x - x_o) + a_2(x - x_o)(x - x_1) \tag{4.8}$$

$$a_0 = y_0 \tag{4.9}$$

$$a_1 = \frac{y_1 - y_0}{x_1 - x_0} \tag{4.10}$$

$$a_2 = \frac{\frac{y_2 - y_1}{x_2 - x_1} - \frac{y_1 - y_0}{x_1 - x_0}}{x_2 - x_0} \tag{4.11}$$

Infeasible, Abnormal, or Outlier Data
The problem of the presence of unrealistic values in the database is common in any experimental work. They could occur due to sensor saturation, electric noise, mechanical vibration, or any error in the operation of the sensor.

The first step is to identify data that are considered infeasible or abnormal. For example, accelerations higher than a certain threshold value ($a > 2.5$ m/s^2) are unrealistic in ground vehicles when sampling at 1 Hz [23, 56]. Oxygen concentrations surpassing 21% are unrealistic in tailpipe emissions [31] because tailpipe gases are byproducts of combustion processes. Therefore, in the absence of combustion (worst condition), the maximum oxygen concentration is equal to the oxygen concentration in the air. When such data is identified, the entire data point is disregarded, or an interpolated value replaces the specific value.

An outlier is an observation that lies at an abnormal distance from other values in a random sampling process from a population.

When the variable is assumed to be normally distributed, two common tests are available on typical statistical software to detect single outliers: the Grubbs test, recommended for relatively large sample sizes ($n > 30$), and the Dixon test, recommended for small samples.

Fig. 4.6 Illustration of the methods to identify atypical values in the experimental data obtained monitoring the operational variables of vehicles

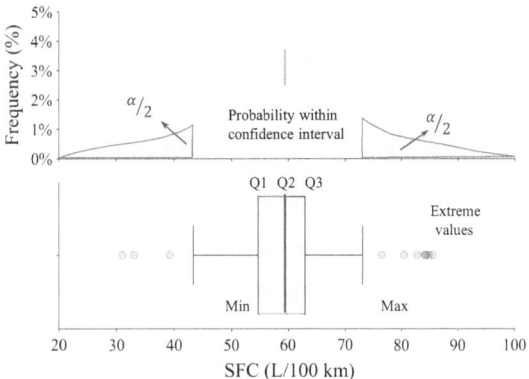

Outliers can also be identified as those values outside the $(1 - \alpha)$ interval around the mean of a given significance level α, which can be calculated through Equation 4.12. In this equation, $\bar{x}$ is the sample mean value, $z_{\alpha/2}$ is the critical value from the standard normal distribution for the given α, and s is the sample standard deviation. α values below 0.01 or equivalently $z_{\alpha/2}$ values greater than 2.3 are recommended.

$$I_{1-\alpha} = \bar{x} \pm z_{\alpha/2} s \qquad (4.12)$$

Alternatively, extreme percentiles can be considered outliers for variables with an asymmetric or unknown distribution. In statistics, a percentile is a term that describes how a data value compares to other values from the same set. In practice, the population of experimental values is sorted from the lowest to the highest value, and the x-percentile corresponds to the value below which an x% of the data falls. For example, values below the 1 percentile and above the 99-percentile can be considered outliers (Fig. 4.6).

Occasionally, unrealistic values occur when derived from discretized experimental data. For example, in the determination of driving patterns, acceleration is obtained as the time derivative of the experimentally measured time series of speed. In this case, we recommend replacing the first-order approximation (Eq. 4.13) of the derivative with a four-order approximation (Eq. 4.14) of the time derivative to mitigate the occurrence of these unrealistic values. It is highlighted that Eq. 4.14 applies only when the time intervals Δt between the consecutive data of speeds (V_i) are the same. In these equations a_i is the estimated acceleration a time i (t_i).

$$a_i = \frac{V_i - V_{i-1}}{t_i - t_{i-1}} \qquad (4.13)$$

$$a_i = \frac{-V_{i+2} + 8V_{i+1} - 8V_{i-1} + V_{i-2}}{12\Delta t} \qquad (4.14)$$

Similarly, problems of unrealistic values occur with variables that are obtained as the ratio of two or more variables. For example, the specific fuel consumption (SFC), is expressed in liter per kilometer traveled (L/km). Problems occur when the time steps are too short for the vehicle to move over a relevant distance (d ~0 km) or the vehicle is near idling. Whenever the mean value of SFC over a trip is searched, we recommend adding the total fuel consumption and the total distance traveled and then reporting the ratio of these variables in opposition to reporting the average of the instant SFCs.

Data Synchronization
Time alignment problems arise when the gathered data is sourced from different sensors or instrumental systems. For example, when measuring tailpipe emissions using data from the ECU and gas analyzers. Even though the clocks of the individual dataloggers or sensors are chronologically synchronized at the start of the tests (all clocks report at the same time) and the data is obtained simultaneously, the resulting data are not synchronized. The time alignment problem arises due to differences in: (i) the physical location of the sensor (Fig. 4.7); (ii) the sensors' time response, (iii) pollutant diffusivities, and (iv) differences in the physical operating principle of each sensor [31, 63].

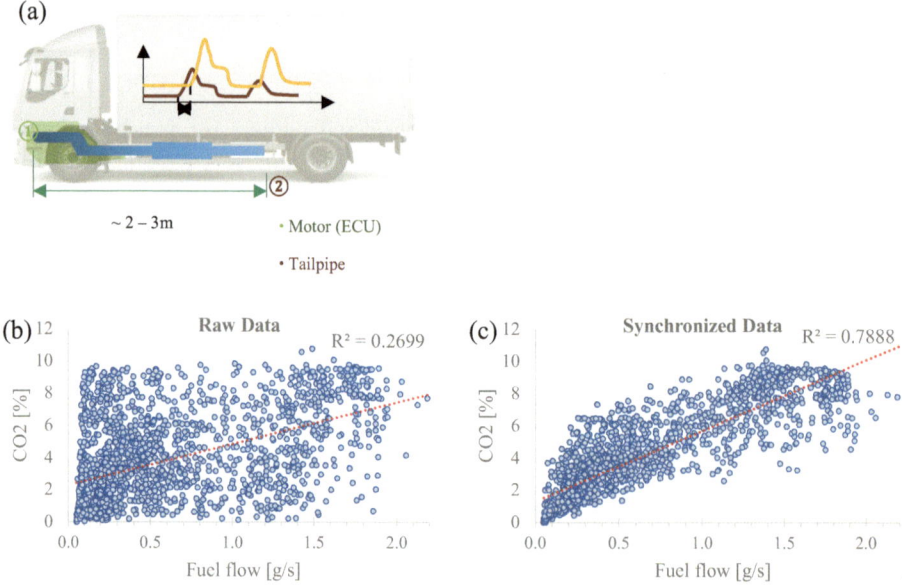

Fig. 4.7. Data synchronization. **a** Illustration of the a-synchronization of the data when several sensors or instrumental systems are used to monitor the operation of vehicles. The correlation between CO_2 emission and fuel consumption rate is illustrated in the data **b** before and **c** after synchronization.

Over the last decade, the implementation of different methods and strategies to synchronize the signals obtained by various devices has been reported. The following synchronization methods are recurrent under the context of monitoring vehicles by on-road tests [31, 63].

Datalogger Alignment

Instead of going through a laborious synchronization process, some studies focus on verifying the time-aligning of the dataloggers or each device used to record the data sampled. It means considering only the data from a certain time recorded in the internal clock of each device of interest [87, 88]. As mentioned, this step is mandatory but does not guarantee time alignment of data.

By Events

This is the most widely used method for synchronizing data gathered by PEMS through on-road tests. It synchronizes two signals of interest using significant changes or peaks in them as criteria. In this method, the first significant change in each of the variables being synchronized is located and matched in time. Usually, both signals are synchronized at the first drastic change or at the first peak. In some cases, these events of significant change or peak values are intentionally generated by accelerating the vehicle before or after each test. This method is usually performed between physically related signals, such as CO_2 or CO concentration with exhaust flow rate, vehicle speed, or engine RPM [89–94].

Recurrent

In this method, any of the previous options are applied at fixed time intervals (every 30–40 s, Cheng et al. [95]).

R^2 Maximization

In this case, variables that physically should be correlated, such as CO_2 emissions and fuel consumption, are used for synchronization. A correlation analysis between these two variables is performed, and the coefficient of determination R^2 is used as a metric to evaluate the degree of correlation. Iterative time shifting (Δt_s) on one of the variables is performed until the resulting R^2 is maximized. We recommend the use of the nonlinear algorithm GRG (Generalized Reduced Gradient) [96] to find the time shifting that maximizes R^2 automatically [97–100].

Dynamic R^2 Maximization

This new method proposed by Giraldo et al. (2024) also uses the maximization of R^2 between two physically related variables to synchronize them. However, due to the nature of the data obtained by PEMS, it considers the lag between both signals as a lag with a constant component (Δt_s) plus a variable component. The latter variable component is directly related to the dynamics of the gases inside the exhaust pipe since it is calculated from variables such as the total exhaust gas flow. This approach allows a dynamic adjustment of the offset between both signals, responding better to the different operating modes of the engine and, therefore, of the vehicle.

$$\Delta t = \Delta t_s + \frac{C_1}{\lambda \dot{m}_f} \tag{4.15}$$

Drifting Effects

Potential drifting issues with CO and NOx sensors, as well as "zero speed drift" effects from GPS, are also addressed in data preprocessing [23, 63]. The zero-speed drift effect typically occurs when a vehicle remains stationary for an extended period, and GPS data loggers record very small speed values. In such cases, the recorded data is often replaced by zero speed.

4.10   Data Analysis Techniques

Finally, the obtained datasets, free from errors, are analyzed to accomplish the established objective described in Sect. 4.1. Several data analysis techniques are available under the context of driving patterns. Next, we briefly describe some of them. We recommend consulting the references indicated for a full description of each technique. Some of these techniques will be described and used in the next chapter.

Classification Methods

There are many techniques to classify data to tag and filter for a posterior analysis. Some machine learning techniques methods include Support Vector Machines (SVM), AdaBoost, and Random Forest. These methods classify driver behavior based on various categories, such as fuel consumption, steering stability, and other relevant factors [101]. The SVM is a supervised machine learning method for classification and regression tasks. It can classify driving behaviors based on various features like speed, acceleration, and braking. An algorithm that combines multiple weak learners to create a strong one is AdaBoost. Another algorithm that uses different decision trees to create a more accurate model is the Random Forest. It could be used to not only classify but also predict emissions and fuel consumption. The regression analysis can predict the class of a given data point based on its features. Unsupervised machine learning techniques for data analysis, such as clustering, have also been explored.

Cluster Analysis

Clustering is an unsupervised machine-learning technique. It consists of grouping data points into clusters based on their similarities and identifying patterns and relationships in the data. Cluster analysis can help, for example, to identify driving behaviors for describing and making predictions. Typically, the researchers have used clustering methods like K-means and X-means to analyze driving cycles [102–104]. Specific clustering methods, such as X-Means, demonstrate the ability to find a fair tradeoff between accuracy and efficiency, adding flexibility to the analysis process. A multi-clustering algorithm framework has been proposed to solve driving cycle prediction problems based on unbalanced data sets, demonstrating the effectiveness of multi-clustering in characterizing operating characteristics [105].

Markov Chains
A probabilistic model is used to describe sequences of events in a system. They have been used in driving cycle analysis for various purposes, such as driving cycle synthesis, development, and evaluation [106]. Markov chains are used to construct driving cycles that reflect driver behavior and local traffic characteristics. The DCs are often described regarding vehicle velocity time profiles and can be used for power management design. Markov Chains can artificially generate equivalent DC. It is possible to iterate and construct driving cycles based on the real driving profiles, or state-based design, where the driving states represent different phases of a driving cycle, such as acceleration, cruising, and braking [107].

Monte Carlo Simulation
Estimating the probability distribution of a given variable by generating a large number of random samples can be done using Monte Carlo Simulation. This method is particularly useful for problems that involve uncertainty and variability. Exploring multiple scenarios and their likelihood of occurrence is possible, giving the user more confidence to make decisions. For example, Monte Carlo Simulation is used in driving cycle analysis for various purposes to estimate vehicle emissions. In this case, it models the uncertainty and variability in driving behavior and traffic conditions. The Monte Carlo method has been used with Markov chains to generate driving cycles.

Principal Component Analysis (PCA)
The PCA is used to reduce the dimensionality of a dataset while retaining most of the variability present in the data. It is very useful when the dataset has high dimensions. This analysis technique identifies the most important variables that contribute to the variability of the data. PCA can be used in data acquisition through sensors. In that case, it is used to identify the most important variables, such as speed, acceleration, or fuel consumption, that could affect driving behavior [108]. Also, it is possible to identify different patterns in driving behavior, like aggressive driving patterns, such as abrupt changes in acceleration and hard braking, or fuel-efficient driving patterns, such as maintaining a steady speed and avoiding sudden changes. Using PCA allows identifying driving behavior patterns

associated with specific road conditions or environments and distinguishinges between urban versus rural driving [109].

Neural Networks (NN)

Neural networks are a learning process that uses interconnected nodes or neurons in a layered structure that resembles the human brain and processes large sets of labeled or unlabeled data. They are organized into two main types: feed-forward and recurrent neural networks (RNNs). Feed-forward neural networks processes data in one direction, from the input node to the output note, while RNNs have feedback loops, making them particularly useful for processing time-series data. It can analyze data and identify complex patterns as driving cycles. Neural networks are particularly useful for anomaly detection or differentiating aggressive and defensive driving behavior [110].

4.11   Concluding Remarks

This chapter described the design and implementation process of monitoring campaigns oriented toward obtaining driving patterns.

During the design of monitoring campaigns, the relevance of clearly stating the objectives sought with the monitoring campaign was highlighted. Most of the time, those objectives require the monitoring of vehicle speed. However, the advancement of technology has enabled the monitoring of other vehicle operation variables where fuel or energy consumption plays a dominant role. Currently, there is a demand for the measurement of tailpipe emissions.

During the implementation of the monitoring campaign, we underlined the relevance of monitoring the vehicles under their normal everyday conditions of operation, with the vehicles being driven by their regular users without imposing restrictions on the roads used. We also underlined that nowadays, telemetry systems can monitor a large number of vehicles during long periods at reduced costs. We discussed alternatives to quantify the minimum number of vehicles and the duration of the monitoring campaign needed to assure a given level of representativeness of the monitoring campaign. We concluded that quantifying the representativeness is still an open question and that, in practice, the monitoring campaign should be carried out with the maximum number of vehicles possible with the limitations imposed by the available economic and time resources.

Finally, during the postprocessing of the data gathered, we presented alternatives to identify errors in the data set and listed some common techniques for analyzing those datasets under the context of driving patterns. These datasets are the input data to the methods designed to describe driving patterns which were explained in Chap. 3. They are also the input data for the methodologies used to construct driving cycles, which will be described in the next chapter.

References

1. Berzi, L., Delogu, M., Pierini, M.: Development of driving cycles for electric vehicles in the context of the city of Florence. Transp. Res. D Transp. Environ. **47**, 299–322 (2016). https://doi.org/10.1016/j.trd.2016.05.010

2. Lipar, P., Strnad, I., Česnik, M., Maher, T.: Development of urban driving cycle with GPS data post processing. PROMET Traffic Transp. **28**, 353–364 (2016). https://doi.org/10.7307/ptt.v28i4.1916

3. Lee, T.-K., Filipi, Z.: Real-World Driving Pattern Recognition for Adaptive HEV Supervisory Control: Based on Representative Driving Cycles in Midwestern US. 16 April 2012

4. Shen, P., Zhao, Z., Li, J., Zhan, X.: Development of a typical driving cycle for an intra-city hybrid electric bus with a fixed route. Transp. Res. D Transp. Environ. **59**, 346–360 (2018). https://doi.org/10.1016/j.trd.2018.01.032

5. Brady, J., O'Mahony, M.: Development of a driving cycle to evaluate the energy economy of electric vehicles in urban areas. Appl. Energy **177**, 165–178 (2016). https://doi.org/10.1016/j.apenergy.2016.05.094

6. Arun, N.H., Mahesh, S., Ramadurai, G., Shiva Nagendra, S.M.: Development of driving cycles for passenger cars and motorcycles in Chennai, India. Sustain Cities Soc **32**, 508–512 (2017). https://doi.org/10.1016/j.scs.2017.05.001

7. Anida, I.N., Ismail, I.S., Norbakyah, J.S., Atiq, W.H., Salisa, A.R.: Characterisation and development of driving cycle for work route in Kuala Terengganu. Int. J. Autom. Mech. Eng. **14**, 4508–4517 (2017). https://doi.org/10.15282/ijame.14.3.2017.9.0356

8. Seedam, A., Satiennam, T., Radpukdee, T., Satiennam, W.: Development of an onboard system to measure the on-road driving pattern for developing motorcycle driving cycle in Khon Kaen City, Thailand. IATSS Res. **39**, 79–85 (2015). https://doi.org/10.1016/j.iatssr.2015.05.003

9. Lai, J., Yu, L., Song, G., Guo, P., Chen, X.: Development of city-specific driving cycles for transit buses based on VSP distributions: case of Beijing. J. Transp. Eng. **139**, 749–757 (2013). https://doi.org/10.1061/(ASCE)TE.1943-5436.0000547

10. Nguyen, Y.-L.T., Nghiem, T.-D., Le, A.-T., Bui, N.-D.: Development of the typical driving cycle for buses in Hanoi, Vietnam. J. Air. Waste Manag. Assoc. **69**, 423–437 (2019). https://doi.org/10.1080/10962247.2018.1543736

11. Knez, M., Muneer, T., Jereb, B., Cullinane, K.: The estimation of a driving cycle for Celje and a comparison to other European cities. Sustain. Cities Soc. **11**, 56–60 (2014). https://doi.org/10.1016/j.scs.2013.11.010

12. Ma, R., He, X., Zheng, Y., Zhou, B., Lu, S., Wu, Y.: Real-world driving cycles and energy consumption informed by large-sized vehicle trajectory data. J. Clean. Prod. **223**, 564–574 (2019). https://doi.org/10.1016/j.jclepro.2019.03.002

13. Zhang, J., Wang, Z., Liu, P., Zhang, Z., Li, X., Qu, C.: Driving cycles construction for electric vehicles considering road environment: a case study in Beijing. Appl. Energy **253**, 113514 (2019). https://doi.org/10.1016/j.apenergy.2019.113514

14. Günther, R., Wenzel, T., Wegner, M., Rettig, R.: Big data driven dynamic driving cycle development for busses in urban public transportation. Transp. Res. D Transp. Environ. **51**, 276–289 (2017). https://doi.org/10.1016/j.trd.2017.01.009

15. Li, X., Zhang, Q., Peng, Z., Wang, A., Wang, W.: A data-driven two-level clustering model for driving pattern analysis of electric vehicles and a case study. J. Clean. Prod. **206**, 827–837 (2019). https://doi.org/10.1016/j.jclepro.2018.09.184

16. Lee, T.-K., Adornato, B., Filipi, Z.S.: Synthesis of real-world driving cycles and their use for estimating PHEV energy consumption and charging opportunities: case study for midwest/

U.S. IEEE Trans. Veh. Technol. **60**, 4153–4163 (2011). https://doi.org/10.1109/TVT.2011.216 8251

17. Tong, H.Y., Tung, H.D., Hung, W.T., Nguyen, H.V.: Development of driving cycles for motorcycles and light-duty vehicles in Vietnam. Atmos. Environ. **45**, 5191–5199 (2011). https://doi.org/10.1016/j.atmosenv.2011.06.023

18. Kaymaz, H., Korkmaz, H., Erdal, H.: Development of a driving cycle for Istanbul bus rapid transit based on real-world data using stratified sampling method. Transp. Res. D Transp. Environ. **75**, 123–135 (2019). https://doi.org/10.1016/j.trd.2019.08.023

19. Hou, D., Sun, Q., Bao, C., Cheng, X., Guo, H., Zhao, Y.: An all-in-one design method for plug-in hybrid electric buses considering uncertain factor of driving cycles. Appl. Energy **253**, 113499 (2019). https://doi.org/10.1016/j.apenergy.2019.113499

20. Gao, J., Xu, Z., Gao, X.: Control strategy for PHEB based on actual driving cycle with driving style characteristic. J. Control Sci. Eng. **2019**, 1–14 (2019). https://doi.org/10.1155/2019/149 6202

21. Zhao, X., Yu, Q., Ma, J., Wu, Y., Yu, M., Ye, Y.: Development of a representative EV urban driving cycle based on a K-means and SVM hybrid clustering algorithm. J. Adv. Transp. **2018**, 1–18 (2018). https://doi.org/10.1155/2018/1890753

22. Yang, Y., Li, T., Hu, H., Zhang, T., Cai, X., Chen, S., Qiao, F.: Development and emissions performance analysis of local driving cycle for small-sized passenger cars in Nanjing, China. Atmos. Pollut. Res. **10**, 1514–1523 (2019). https://doi.org/10.1016/j.apr.2019.04.009

23. Tong, H.Y.: Development of a driving cycle for a supercapacitor electric bus route in Hong Kong. Sustain. Cities Soc. **48**, 101588 (2019). https://doi.org/10.1016/j.scs.2019.101588

24. Chauhan, B.P., Joshi, G.J., Purnima, P.: Candidate Driving Cycle Construction for Emission Estimation, pp. 85–97 (2020)

25. Geetha, A., Subramani, C.: Development of driving cycle under real world traffic conditions: a case study. Int. J. Electr. Comput. Eng. (IJECE) **9**, 4798 (2019). https://doi.org/10.11591/ijece. v9i6.pp4798-4803

26. Saleh, W., Kumar, R., Kirby, H., Kumar, P.: Real world driving cycle for motorcycles in Edinburgh. Transp. Res. D Transp. Environ. **14**, 326–333 (2009). https://doi.org/10.1016/j.trd.2009. 03.003

27. Zhao, J., Gao, Y., Guo, J., Chu, L.: The creation of a representative driving cycle based on intelligent transportation system (ITS) and a mathematically statistical algorithm: a case study of Changchun (China). Sustain. Cities Soc. **42**, 301–313 (2018). https://doi.org/10.1016/j.scs. 2018.05.031

28. Huertas, J.I., Díaz, J., Cordero, D., Cedillo, K.: A new methodology to determine typical driving cycles for the design of vehicles power trains. Int. J. Interact. Des. Manuf. (IJIDeM) **12**, 319–326 (2018). https://doi.org/10.1007/s12008-017-0379-y

29. Gong, Q., Midlam-Mohler, S., Marano, V., Rizzoni, G.: An iterative Markov chain approach for generating vehicle driving cycles. SAE Int. J. Engines **4** (2011). https://doi.org/10.4271/2011-01-0880

30. He, H., Sun, C., Zhang, X.: A method for identification of driving patterns in hybrid electric vehicles based on a LVQ neural network. Energies (Basel) **5**, 3363–3380 (2012). https://doi.org/10.3390/en5093363

31. Quirama, L.F., Giraldo, M., Huertas, J.I., Jaller, M.: Driving cycles that reproduce driving patterns, energy consumptions and tailpipe emissions. Transp. Res. D Transp. Environ. **82**, 102294 (2020). https://doi.org/10.1016/j.trd.2020.102294

32. Huertas, J., Quirama, L., Giraldo, M., Díaz, J.: Comparison of three methods for constructing real driving cycles. Energies (Basel) **12**, 665 (2019). https://doi.org/10.3390/en12040665

33. Shi, Q., Qiu, D., He, L., Wu, B., Li, Y.: Support vector machine-based driving cycle recognition for dynamic equivalent fuel consumption minimization strategy with hybrid electric vehicle. Adv. Mech. Eng. **10**, 168781401881102 (2018). https://doi.org/10.1177/1687814018811020

34. Huertas, J., Giraldo, M., Quirama, L., Díaz, J.: Driving cycles based on fuel consumption. Energies (Basel) **11**, 3064 (2018). https://doi.org/10.3390/en11113064

35. Mongkonlerdmanee, S., Koetniyom, S.: Development of a realistic driving cycle using time series clustering technique for buses: Thailand case study. Eng. J. **23**, 49–65 (2019). https://doi.org/10.4186/ej.2019.23.4.49

36. Chen, Z., Li, L., Yan, B., Yang, C., Marina Martinez, C., Cao, D.: Multimode energy management for plug-in hybrid electric buses based on driving cycles prediction. IEEE Trans. Intell. Transp. Syst. **17**, 2811–2821 (2016). https://doi.org/10.1109/TITS.2016.2527244

37. Shi, Q., Liu, B., Guan, Q., He, L., Qiu, D.: A genetic ant colony algorithm-based driving cycle generation approach for testing driving range of battery electric vehicle. Adv. Mech. Eng. **12**, 168781401990105 (2020). https://doi.org/10.1177/1687814019901054

38. Hu, J., Niu, X., Jiang, X., Zu, G.: Energy management strategy based on driving pattern recognition for a dual-motor battery electric vehicle. Int. J. Energy Res. **43**, 3346–3364 (2019). https://doi.org/10.1002/er.4474

39. Ho, S.-H., Wong, Y.-D., Chang, V.W.-C.: Developing Singapore driving cycle for passenger cars to estimate fuel consumption and vehicular emissions. Atmos. Environ. **97**, 353–362 (2014). https://doi.org/10.1016/j.atmosenv.2014.08.042

40. Gao, X., Zhang, B., Xiong, X., Dong, M., Li, H.: Construction and analysis of the Dalian driving cycle. Int. J. Control Autom. **8**, 363–368 (2015). https://doi.org/10.14257/ijca.2015.8.6.35

41. Nesamani, K.S., Subramanian, K.P.: Development of a driving cycle for intra-city buses in Chennai, India. Atmos. Environ. **45**, 5469–5476 (2011). https://doi.org/10.1016/j.atmosenv.2011.06.067

42. Liu, B., Shi, Q., He, L., Qiu, D.: A study on the construction of Hefei urban driving cycle for passenger vehicle. IFAC-PapersOnLine **51**, 854–858 (2018). https://doi.org/10.1016/j.ifacol.2018.10.100

43. Schüller, M., Tewiele, S., Bruckmann, T., Schramm, D.: Evaluation of alternative drive systems based on driving patterns comparing Germany, China and Malaysia. Int. J. Autom. Mech. Eng. **14**, 3985–3997 (2017). https://doi.org/10.15282/ijame.14.1.2017.13.0323

44. Mayakuntla, S.K., Verma, A.: A novel methodology for construction of driving cycles for Indian cities. Transp. Res. D Transp. Environ. **65**, 725–735 (2018). https://doi.org/10.1016/j.trd.2018.10.013

45. Lin, C., Zhao, L., Cheng, X., Wang, W.: A DCT-based driving cycle generation method and its application for electric vehicles. Math. Probl. Eng. **2015**, 1–13 (2015). https://doi.org/10.1155/2015/178902

46. Tamsanya, S., Chungpaibulpatana, S., Limmeechokchai, B.: Development of a driving cycle for the measurement of fuel consumption and exhaust emissions of automobiles in Bangkok during peak periods. Int. J. Automot. Technol. **10**, 251–264 (2009). https://doi.org/10.1007/s12239-009-0030-4

47. Han, D.S., Choi, N.W., Cho, S.L., Yang, J.S., Kim, K.S., Yoo, W.S., Jeon, C.H.: Characterization of driving patterns and development of a driving cycle in a military area. Transp. Res. D Transp. Environ. **17**, 519–524 (2012). https://doi.org/10.1016/j.trd.2012.06.004

48. Zhao, X., Ma, J., Wang, S., Ye, Y., Wu, Y., Yu, M.: Developing an electric vehicle urban driving cycle to study differences in energy consumption. Environ. Sci. Pollut. Res. **26**, 13839–13853 (2019). https://doi.org/10.1007/s11356-018-3541-6

49. Galgamuwa, U., Perera, L., Bandara, S.: Developing a general methodology for driving cycle construction: comparison of various established driving cycles in the world to propose a general approach. J. Transp. Technol. **05**, 191–203 (2015). https://doi.org/10.4236/jtts.2015.54018
50. Paraphantakul, C.: Review of worldwide road classification systems. Eng. J. **23** (2014)
51. United Nations - Treaty Series: European Agreement on Main International Traffic Arteries (AGR) (with Annexes and List of Roads). Concluded at Geneva on 15 November 1975; Geneva (1975)
52. Department of Transport - UK Guidance on Road Classification and the Primary Route Network. Statutory Guidance. https://treaties.un.org/doc/Publication/UNTS/Volume%201302/volume-1302-I-21618-English.pdf. Accessed on 7 January 2024
53. Asian Development Bank PRC Road Classification Information. Yunnan Pu'er Regional Integrated Road Network Development Project (RRP PRC 46040). https://www.adb.org/sites/default/files/linked-documents/46040-003-sd-04.pdf. Accessed on 7 January 2024
54. Thurston Regional Planning Council: What Moves You – 2045 Regional Transportation Plan (RTP). Appendix K Level of Service Standard and Measurements; (2020)
55. Ashtari, A., Bibeau, E., Shahidinejad, S.: Using large driving record samples and a stochastic approach for real-world driving cycle construction: Winnipeg driving cycle. Transp. Sci. **48**, 170–183 (2014). https://doi.org/10.1287/trsc.1120.0447
56. Peng, Y., Zhuang, Y., Yang, Y.: A driving cycle construction methodology combining k-means clustering and Markov model for urban mixed roads. Proc. Inst. Mech. Eng. Part D: J. Autom. Eng. **234**, 714–724 (2020). https://doi.org/10.1177/0954407019848873
57. Yu, L., Wang, Z., Qiao, F., Qi, Y.: Approach to development and evaluation of driving cycles for classified roads based on vehicle emission characteristics. Transp. Res. Record: J. Transp. Res. Board **2058**, 58–67 (2008). https://doi.org/10.3141/2058-08
58. Hongwen, H., Jinquan, G., Jiankun, P., Huachun, T., Chao, S.: Real-time global driving cycle construction and the application to economy driving pro system in plug-in hybrid electric vehicles. Energy **152**, 95–107 (2018). https://doi.org/10.1016/j.energy.2018.03.061
59. Bishop, J.D.K., Axon, C.J., McCulloch, M.D.: A robust, data-driven methodology for real-world driving cycle development. Transp. Res. D Transp. Environ. **17**, 389–397 (2012). https://doi.org/10.1016/j.trd.2012.03.003
60. Jing, Z., Wang, G., Zhang, S., Qiu, C.: Building Tianjin driving cycle based on linear discriminant analysis. Transp. Res. D Transp. Environ. **53**, 78–87 (2017). https://doi.org/10.1016/j.trd.2017.04.005
61. Nouri, P., Morency, C.: Evaluating microtrip definitions for developing driving cycles. Transp. Res. Record: J. Transp. Res. Board **2627**, 86–92 (2017). https://doi.org/10.3141/2627-10
62. Segob, S. de G.: Norma Oficial Mexicana NOM-012-SCT-2-2017, Sobre El Peso y Dimensiones Máximas Con Los Que Pueden Circular Los Vehículos de Autotransporte Que Transitan En Las Vías Generales de Comunicación de Jurisdicción Federal (2017)
63. Huertas, J.I., Quirama, L.F., Giraldo, M.D., Diaz, J.: Comparison of driving cycles obtained by the micro-trips, Markov-chains and MWD-CP methods. Int. J. Sustain. Energy Plan. Manag. **22** (2019). https://doi.org/10.5278/ijsepm.2554
64. Wu, X., Hu, C., Du, J.: Development of a driving cycle for city bus in Harbin of China. Int. J. Electr. Hybrid Veh. **7**, 104 (2015). https://doi.org/10.1504/IJEHV.2015.071063
65. Pouresmaeili, M.A., Aghayan, I., Taghizadeh, S.A.: Development of Mashhad driving cycle for passenger car to model vehicle exhaust emissions calibrated using on-board measurements. Sustain. Cities Soc. **36**, 12–20 (2018). https://doi.org/10.1016/j.scs.2017.09.034
66. Pfriem, M., Gauterin, F.: Development of real-world driving cycles for battery electric vehicles. World Electr. Veh. J. **8**, 14–24 (2016). https://doi.org/10.3390/wevj8010014

67. Zhou, W., Xu, K., Yang, Y., Lu, J.: Driving cycle development for electric vehicle application using principal component analysis and K-means cluster: with the case of Shenyang, China. Energy Procedia **105**, 2831–2836 (2017). https://doi.org/10.1016/j.egypro.2017.03.620

68. Hung, W.T., Tong, H.Y., Lee, C.P., Ha, K., Pao, L.Y.: Development of a practical driving cycle construction methodology: a case study in Hong Kong. Transp. Res. D Transp. Environ. **12**, 115–128 (2007). https://doi.org/10.1016/j.trd.2007.01.002

69. He, H., Guo, J., Zhou, N., Sun, C., Peng, J.: Freeway driving cycle construction based on real-time traffic information and global optimal energy management for plug-in hybrid electric vehicles. Energies (Basel) **10**, 1796 (2017). https://doi.org/10.3390/en10111796

70. Koossalapeerom, T., Satiennam, T., Satiennam, W., Leelapatra, W., Seedam, A., Rakpukdee, T.: Comparative study of real-world driving cycles, energy consumption, and CO_2 emissions of electric and gasoline motorcycles driving in a congested urban corridor. Sustain. Cities Soc. **45**, 619–627 (2019). https://doi.org/10.1016/j.scs.2018.12.031

71. Seers, P., Nachin, G., Glaus, M.: Development of two driving cycles for utility vehicles. Transp. Res. D Transp. Environ. **41**, 377–385 (2015). https://doi.org/10.1016/j.trd.2015.10.013

72. NASA JPL NASA Shuttle Radar Topography Mission Global 3 Arc Second [Data Set]. https://doi.org/10.5067/MEaSUREs/SRTM/SRTMGL3.003

73. Mogro, A.E., Huertas, J.I.: Root causes of the differences in the real-world vehicle emissions between Mexico and the US. Transp. Res. D Transp. Environ. **102**, 103153 (2022). https://doi.org/10.1016/J.TRD.2021.103153

74. Lyu, M., Bao, X., Wang, Y., Matthews, R.: Analysis of emissions from various driving cycles based on real driving measurements obtained in a high-altitude city. **234**, 1563–1571 (2020). https://doi.org/10.1177/0954407019898959

75. Yang, Y., Li, T., Zhang, T., Yu, Q.: Time dimension analysis: comparison of Nanjing local driving cycles in 2009 and 2017. Sustain. Cities Soc. **53**, 101949 (2020). https://doi.org/10.1016/j.scs.2019.101949

76. Ragione, L.D., Meccariello, G.: Statistical analysis on real kinematic urban driving cycles by preliminary use of VSP variable. Int. J. Environ. Sci. **04** (2019)

77. Yuhui, P., Yuan, Z., Huibao, Y.: Development of a representative driving cycle for urban buses based on the K-means cluster method. Cluster Comput. **22**, 6871–6880 (2019). https://doi.org/10.1007/s10586-017-1673-y

78. Galgamuwa, U., Perera, L., Bandara, S.: Development of a driving cycle for Colombo, Sri Lanka: an economical approach for developing countries. J. Adv. Transp. **50**, 1520–1530 (2016). https://doi.org/10.1002/atr.1414

79. Zhang, M., Shi, S., Lin, N., Yue, B.: High-efficiency driving cycle generation using a Markov chain evolution algorithm. IEEE Trans. Veh. Technol. **68**, 1288–1301 (2019). https://doi.org/10.1109/TVT.2018.2887063

80. Wang, Z., Zhang, J., Liu, P., Qu, C., Li, X.: Driving cycle construction for electric vehicles based on Markov chain and monte carlo method: a case study in Beijing. Energy Procedia **158**, 2494–2499 (2019). https://doi.org/10.1016/j.egypro.2019.01.389

81. Tong, H.Y., Hung, W.T.: A framework for developing driving cycles with on-road driving data. Transp. Rev. **30**, 589–615 (2010). https://doi.org/10.1080/01441640903286134

82. Kamble, S.H., Mathew, T.V., Sharma, G.K.: Development of real-world driving cycle: case study of Pune, India. Transp. Res. D Transp. Environ. **14**, 132–140 (2009). https://doi.org/10.1016/j.trd.2008.11.008

83. Shi, S., Lin, N., Zhang, Y., Cheng, J., Huang, C., Liu, L., Lu, B.: Research on Markov property analysis of driving cycles and its application. Transp. Res. D Transp. Environ. **47**, 171–181 (2016). https://doi.org/10.1016/j.trd.2016.05.013

84. Hwa, M.Y., Yu, T.Y.: Development of real-world driving cycles and estimation of emission factors for in-use light-duty gasoline vehicles in urban areas. Environ. Monit. Assess. **186**, 3985–3994 (2014). https://doi.org/10.1007/S10661-014-3673-1/TABLES/5

85. Chauhan, B.P., Joshi, G.J., Purnima, P.: Candidate driving cycle construction for emission estimation. In: Mathew, T.V., Joshi, G.J., Velaga, N.R., Arkatkar, S. (eds.) Proceedings of the Transportation Research, pp. 85–97. Springer Singapore, Singapore (2020)

86. Gong, H., Zou, Y., Yang, Q., Fan, J., Sun, F., Goehlich, D.: Generation of a driving cycle for battery electric vehicles: a case Study of Beijing. Energy **150**, 901–912 (2018). https://doi.org/10.1016/j.energy.2018.02.092

87. Chandrashekar, C., Chatterjee, P., Pawar, D.S.: Estimation of CO_2 and CO emissions from auto-rickshaws in Indian heterogeneous traffic. Transp. Res. D Transp. Environ. **104**, 103202 (2022). https://doi.org/10.1016/j.trd.2022.103202

88. Yu, Q., Yang, Y., Xiong, X., Sun, S., Liu, Y., Wang, Y.: Assessing the impact of multi-dimensional driving behaviors on link-level emissions based on a portable emission measurement system (PEMS). Atmos. Pollut. Res. **12**, 414–424 (2021). https://doi.org/10.1016/j.apr.2020.09.022

89. Gallus, J., Kirchner, U., Vogt, R., Börensen, C., Benter, T.: On-road particle number measurements using a portable emission measurement system (PEMS). Atmos. Environ. **124**, 37–45 (2016). https://doi.org/10.1016/j.atmosenv.2015.11.012

90. Meneguzzer, C., Gastaldi, M., Rossi, R., Gecchele, G., Prati, M.V.: Comparison of exhaust emissions at intersections under traffic signal versus roundabout control using an instrumented vehicle. Transp. Res. Procedia **25**, 1597–1609 (2017). https://doi.org/10.1016/j.trpro.2017.05.204

91. Vojtíšek-Lom, M., Beránek, V., Klír, V., Jindra, P., Pechout, M., Voříšek, T.: On-road and laboratory emissions of NO, NO_2, NH_3, N_2O and CH_4 from late-model EU light utility vehicles: comparison of diesel and CNG. Sci. Total. Environ. **616–617**, 774–784 (2018). https://doi.org/10.1016/j.scitotenv.2017.10.248

92. Sandhu, G.S., Frey, H.C., Bartelt-Hunt, S., Jones, E.: Real-world activity, fuel use, and emissions of heavy-duty compressed natural gas refuse trucks. Sci. Total. Environ. **761**, 143323 (2021). https://doi.org/10.1016/j.scitotenv.2020.143323

93. Pang, K., Zhang, K., Ma, S.: Tailpipe emission characterizations of diesel-fueled Forklifts under real-world operations using a portable emission measurement system. J. Environ. Sci. (China) **100**, 34–42 (2021). https://doi.org/10.1016/j.jes.2020.07.011

94. Dhital, N.B., Wang, S.X., Lee, C.H., Su, J., Tsai, M.Y., Jhou, Y.J., Yang, H.H.: Effects of driving behavior on real-world emissions of particulate matter, gaseous pollutants and particle-bound PAHs for diesel trucks. Environ. Pollut. **286**, 117292 (2021). https://doi.org/10.1016/j.envpol.2021.117292

95. Cheng, Y., He, L., He, W., Zhao, P., Wang, P., Zhao, J., Zhang, K., Zhang, S.: Evaluating On-Board Sensing-Based Nitrogen Oxides (NOX) Emissions from a Heavy-Duty Diesel Truck in China. Atmos Environ **216**, 116908 (2019). https://doi.org/10.1016/j.atmosenv.2019.116908

96. Lasdon, L.S., Waren, A.D., Jain, A., Ratner, M.: Design and testing of a generalized reduced gradient code for nonlinear programming. ACM Trans. Math. Softw. (TOMS) **4**, 34–50 (1978). https://doi.org/10.1145/355769.355773

97. Giraldo, M., Huertas, J.I.: Real emissions, driving patterns and fuel consumption of in-use diesel buses operating at high altitude. Transp. Res. D Transp. Environ. **77**, 21–36 (2019). https://doi.org/10.1016/j.trd.2019.10.004

98. Wang, H., Wu, Y., Zhang, K.M., Zhang, S., Baldauf, R.W., Snow, R., Deshmukh, P., Zheng, X., He, L., Hao, J.: Evaluating mobile monitoring of on-road emission factors by comparing

concurrent PEMS measurements. Sci. Total Environ. **736** (2020). https://doi.org/10.1016/j.sci totenv.2020.139507

99. Quirama, L.F., Giraldo, M., Huertas, J.I., Tibaquirá, J.E., Cordero-Moreno, D.: Main characteristic parameters to describe driving patterns and construct driving cycles. Transp. Res. D Transp. Environ. **97** (2021). https://doi.org/10.1016/j.trd.2021.102959

100. He, L., You, Y., Zheng, X., Zhang, S., Li, Z., Zhang, Z., Wu, Y., Hao, J.: The impacts from cold start and road grade on real-world emissions and fuel consumption of gasoline, diesel and hybrid-electric light-duty passenger vehicles. Sci. Total Environ. **851** (2022). https://doi.org/ 10.1016/j.scitotenv.2022.158045

101. kumar, R., Jain, A.: Driving behavior analysis and classification by vehicle OBD data using machine learning. J. Supercomput. **79**, 18800–18819 (2023). https://doi.org/10.1007/S11227-023-05364-3/FIGURES/5

102. Fotouhi, A., Montazeri-Gh, M.: Tehran driving cycle development using the K-means clustering method. Sci. Iranica **20**, 286–293 (2013). https://doi.org/10.1016/j.scient.2013.04.001

103. Salameh, M., Brown, I.P., Krishnamurthy, M.: Driving cycle analysis methods using data clustering for machine design optimization. In: ITEC 2019—2019 IEEE Transportation Electrification Conference and Expo (2019). https://doi.org/10.1109/ITEC.2019.8790523

104. Salameh, M., Brown, I.P., Krishnamurthy, M.: Fundamental evaluation of data clustering approaches for driving cycle-based machine design optimization. IEEE Trans. Transp. Electrif. **5**, 1395–1405 (2019). https://doi.org/10.1109/TTE.2019.2950869

105. Wu, Y., Zhang, W., Zhang, L., Qiao, Y., Yang, J., Cheng, C.: A multi-clustering algorithm to solve driving cycle prediction problems based on unbalanced data sets: a Chinese case study. Sensors **20**, 2448 (2020). https://doi.org/10.3390/S20092448

106. Nyberg, P., Frisk, E., Nielsen, L.: Generation of equivalent driving cycles using Markov chains and mean tractive force components. IFAC Proc. Vol. **47**, 8787–8792 (2014). https://doi.org/ 10.3182/20140824-6-ZA-1003.02239

107. Zähringer, M., Kalt, S., Lienkamp, M.: Compressed driving cycles using Markov chains for vehicle powertrain design. World Electr. Veh. J. **11**, 52 (2020). https://doi.org/10.3390/WEV J11030052

108. Reddy, G.T., Reddy, M.P.K., Lakshmanna, K., Kaluri, R., Rajput, D.S., Srivastava, G., Baker, T.: Analysis of dimensionality reduction techniques on big data. IEEE Access **8**, 54776–54788 (2020). https://doi.org/10.1109/ACCESS.2020.2980942

109. Castignani, G., Chandelle, F., Jafarnejad, S., Scherer, I.: Evaluation of a driver's compatibility with electric, plug-in hybrid, and hybrid vehicles based on mobility patterns analytics. ELIV **2021**, 489–500 (2021). https://doi.org/10.51202/9783181023846-489

110. Chen, Z., Yang, C., Fang, S.: A convolutional neural network-based driving cycle prediction method for plug-in hybrid electric vehicles with bus route. IEEE Access **8**, 3255–3264 (2020). https://doi.org/10.1109/ACCESS.2019.2960771

Driving Cycle Construction Methods

5

Michael Giraldo, Luis Felipe Quirama, and José I. Huertas

Abstract

In this chapter, we describe the methods most commonly employed by researchers to construct DCs. The primary challenge in these methods is ensuring that the resulting DC accurately represents the driving pattern of the region or application under study. Assuming that the data collected during the monitoring campaign is both sufficient in quantity and high in quality, the representativeness of the resulting DC is directly influenced by the chosen DC construction method.

5.1   A Glance of the Construction Methods

The selection of the DC construction method can be driven by the purpose for which the DC is needed [1–3]. However, in practice, it has been driven by the researchers' preferences. DC construction methods can be divided into stochastic and deterministic (Table 5.1). Next, we describe each method. Figure 5.1 illustrates the algorithms typically

M. Giraldo
Industry, Materials, and Energy Area, School of Applied Sciences and Engineering, Universidad EAFIT, 050030 Medellín, Colombia
e-mail: mgiral36@eafit.edu.co

L. F. Quirama
Sustainable Mobility Unit, United Nations Environment Programme, 660003 Panama City, Panama
e-mail: luis.felipe@un.org

J. I. Huertas (✉)
Sustainable Energy Research Group, Tecnologico de Monterrey, Monterrey 64849, México
e-mail: jhuertas@tec.mx

© The Author(s), under exclusive license to Springer Nature Switzerland AG 2025
J. I. Huertas (ed.), *Fundamentals of Driving Patterns and Driving Cycles*, Synthesis Lectures on Mechanical Engineering, https://doi.org/10.1007/978-3-031-76863-7_5

Table 5.1 The DC construction methods most commoly used or reported in the literature

	Method	Pros	Cons	References
Stochastic	Micro-trips	• Most frequently used • Fuel consumption and tailpipe emissions can be included	Not repeatable (different DCs are produced when the method is replicated with the same input data)	[4, 6–14] [15, 16]
	Markov chain	Driving conditions are represented through probabilities of occurrence of each mode or state	• Not repeatable • Large computational effort • Could include unrealistic accelerations within the DC • Fuel consumption and emissions have not been included	[4, 9, 17–25] [26–31]
	Energy-based micro-trip	Include the reproduction of energy consumption and emissions	Not repeatable	[32]
Deterministic	Modal-based	• Can represent physical features of the road/ driving (i.e., expressways) • Reproducible and repeatable	• Difficult segment assembly • Could include unrealistic accelerations within the DC	[33–39]
	Trip-based	• Fuel consumption and emission determination included • Reproducible and repeatable	DC duration could be an issue	[40–42]

used to implement them. A comparison of different DC construction methods can also be found in [2, 4, 5].

5.2   The Micro-Trips Method

The micro-trip (MT) method is by far the most recurrent method used for constructing DCs. However, authors have multiple variations of it. Below, we describe the most frequent and recommended features to be included in its implementation.

Construction of MT Database
The input data for this method is the large set of trips of different time durations and distances collected during the monitoring campaign carried out to grasp the driving pattern. In this method, each trip of that database is partitioned into micro-trips (Fig. 5.2). They are a speed versus time segment with initial and final speeds equal to zero, which could include an idling time at the beginning or the end. The collection of the resulting segments

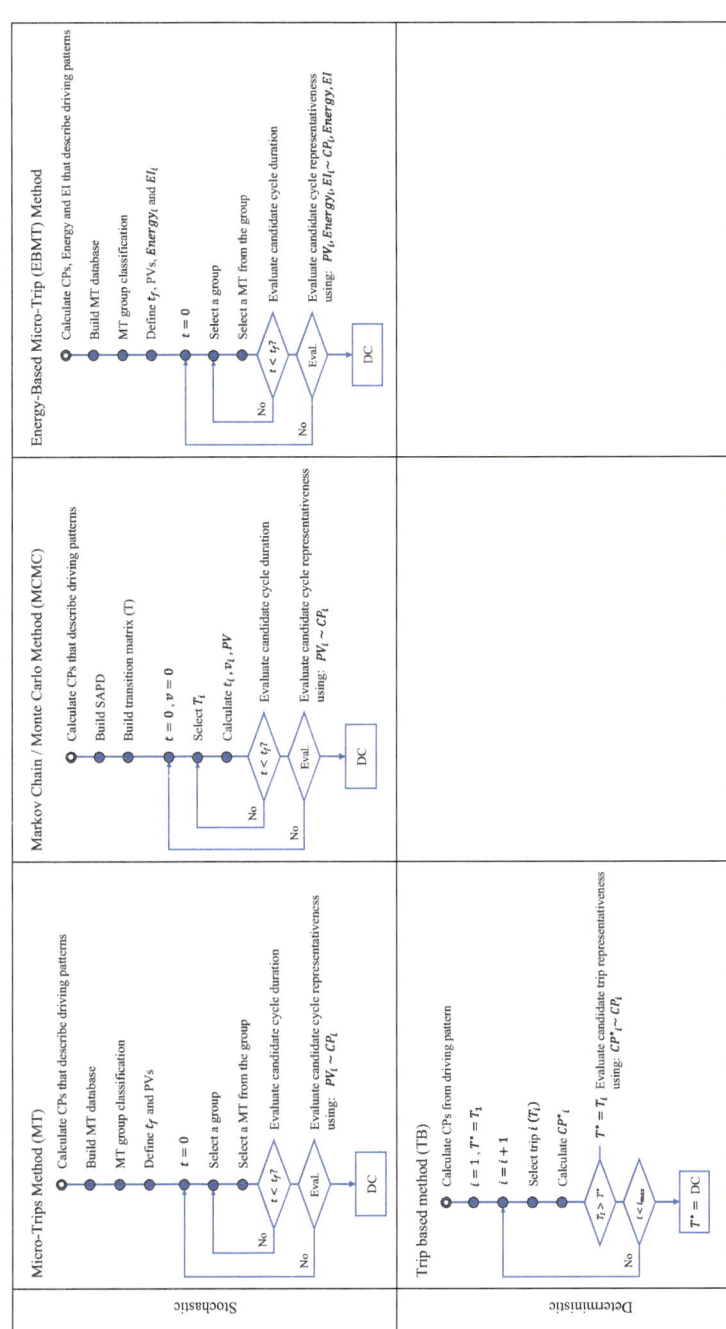

Fig. 5.1 Algorithms to implement the most frequent DC construction methods. CPs: Characteristics parameters of the driving pattern; CPs*: characteristic parameter of the DC; EI: Emission index

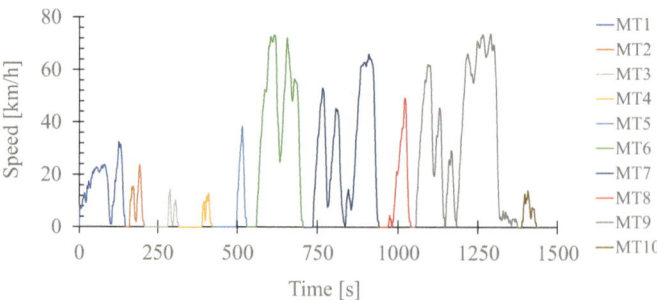

Fig. 5.2 Illustration of the process of splitting one trip into several micro-trips

forms a new database of micro-trips. A few studies split the trips into segments based on other criteria, sometimes called "kinematic segments," such as operational modes [3]. This last option could be used when stops or zero speeds are not frequent, like in the case of very long single trips or highway driving patterns. The main drawback of this option is that additional processing is needed to guarantee a physically feasible matching or assembly of micro-segments since the initial and final speeds of the resulting micro-segments could be of any value [3]. Therefore, unrealistic vehicle accelerations could artificially result in the upcoming steps when micro-segments are spliced together.

Clustering

A clustering step is optionally applied for gathering micro-trips with similar characteristics. Authors group them according to the driving style (eco-friendly, normal, and aggressive), the road type (low speed, highway), traffic condition (congested, typical, free), or just according to one or up to three CPs. This last option is the most frequent. The most recurrent CPs used are mean speed and mean positive acceleration (Fig. 5.3). Usually, researchers use between 5 and 10 clusters per CP. Alternatively, optimization algorithms can be used to identify the minimum number of clusters. The micro-trips can also be grouped according to a third criterion, such as the level of similitude. The algorithms most frequently used to perform this clustering step are the centroids linkage method and K-means [9, 15, 19, 30, 33, 34, 38, 43–46].

Assembling a Candidate DC

Then MTs are selected and spliced together to form a candidate DC. Several techniques are used for this selection. Some authors make sure that each cluster is present in the candidate DC by picking one MT from each cluster. Others select the cluster according to each cluster's probability distribution of occurrences, i.e., proportionally to the population of MTs in each cluster. Within each cluster, authors select arbitrarily any of the MTs or the one with the average characteristics within that cluster. Combinations of these alternatives are possible. To construct representative DC, we recommend randomly selecting the MT

Fig. 5.3 Illustration of the micro-trip clustering process when the means of speed and positive acceleration are used as criteria

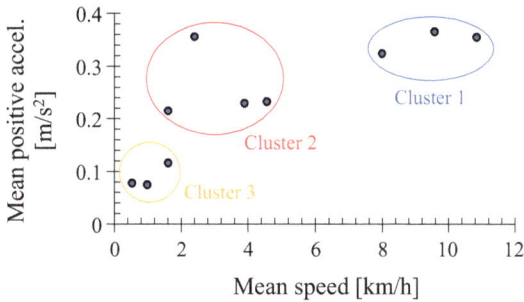

within each cluster and selecting the cluster according to its frequency of occurrence. This recommendation is equal to eliminating the clustering step from the construction process and randomly selecting the MT from the whole database. The assembly process continues progressively until the candidate DC reaches a minimum duration (or traveled distance). Section 5.4 will address the appropriate duration of DCs.

Criteria for Assessing the Representativeness of the Candidate DC
The representativeness of a candidate DC is evaluated based on how well it represents the targeted driving pattern. However, there is a lack of consensus regarding the CPs that should be used during this assessment. This issue will be addressed in Sect. 6.2. For consistency, the criteria used to evaluate the representativeness of the candidate DC should align with the criteria employed in the clustering method. Typically, 2 or 3 CPs are used as assessment parameters, with mean speed, percentage of idling time, and mean positive acceleration being the most used. Conceptually, this approach implies that these two or three CPs can adequately describe the driven pattern. Therefore, for a candidate DC to be considered representative, it should have the same mean speed, percentage of idling time, and mean positive acceleration as the driving pattern. The representativeness of the candidate DC is assessed through Eq. 5.1. *RDi* represents the relative difference between the *CPi** that describes the candidate DC and the *CPi* that describes the driving pattern. The index *i* denotes any of the CPs used as assessment parameters. It is important to note that *RDi* can range from 0 to infinity. Previous studies have used specific threshold values to determine the acceptability of the candidate DC based on the calculated RD_i values [3, 7, 13, 18, 47].

$$RD_i = \frac{\left|CP_i^* - CP_i\right|}{CP_i}$$ (5.1)

When the authors consider that the SAPD describes the driving pattern, they ensure that a candidate DC exhibits an SAPD* similar to the SAPD of the driving pattern. In this case, they use Eq. 5.2 to assess that representativeness.

$$RD_{SAPD} = \frac{\sum_{i=1}^{m} \sum_{j=1}^{r} \left| P_{ij}^* - P_{ij} \right|}{2} \qquad (5.2)$$

where P_{ij}^* is the probability that the vehicle travels at speed i and acceleration j according to the candidate DC, and P_{ij} is the same variable for the driving pattern. r and m are the bins used to discretize the speed and acceleration, respectively. The maximum value that can reach the absolute difference between P_{ij}^* and P_{ij} is 2; therefore, RD_{SAPD} ranges between 0 and 1. Usually, 10^{-4} is used as the threshold value for acceptance [40]. Previous criteria can be extended for the cases when the driven pattern is described by other alternatives, such as the ones described in Sect. 2.7.

End Condition
When the assessment criteria for representatives are unsatisfied, the MT method restarts and selects a new group of micro-trips. The first candidate DC that satisfies the assessment criteria becomes the "representative" DC, and the algorithm ends.

This method is reproducible but not repeatable, as the DC obtained changes each time the method is replicated using the same database. This feature is not a technical issue if any of the obtained DCs meet the objective established in the first step. Aiming to mitigate the problem of lack of repeatability of this method, we suggest repeating the MT method many times (~1000) and keeping the DC with the minimum average RDi. In this equation, n represents the total number of repetitions, and j denotes the iteration number.

$$ARD_j = \frac{\sum_{j=1}^{n} \left| CP_{i,j}^* - CP_i \right|}{nCP_i} \qquad (5.3)$$

Among the DCs that were constructed using the MT method are LA92, Singapore, Hong Kong, and Bangkok DCs [6, 47–49].

5.3   Energy-Based Micro-Trip (EBMT) Method

Considering that DCs are mainly used for quantifying fuel consumption and tailpipe emissions, the Energy-Based Micro-Trip (EBMT) method is an MT method that seeks to have DCs that are not only representative of the driving patterns of the region but also in terms of energy consumption and vehicle emissions.

Tong and Hung addressed energy consumption and vehicle emissions when constructing driving cycles. They recommended simultaneously monitoring driving patterns, energy consumption, and vehicle emissions [2]. Xie et al. used a fuel consumption estimation to cluster the micro-trips [50]. Huertas and Coello used a set of CPs to estimate vehicle-specific fuel consumption (SFC) [51]. Later, Huertas et al. used fuel consumption and emissions data to propose the deterministic method called the Fuel-based Method, where a trip is selected as a representative DC if its fuel consumption is close to the average fuel consumption of the monitored trips [40]. The effectiveness of the fuel-based method in producing representative driving cycles concerning traditional methods such as MTs and Markov Chain Monte Carlo was analyzed in [40].

The EBMT is a stochastic method based on the MT method, where the representativeness of the candidate DC is evaluated using as criteria the relative difference (RDi, Eq. 5.1) of the specific energy consumption, average speed, percentage of idling time, and average positive acceleration. Therefore, the EBMT method requires that during the monitoring campaign of the operative vehicle conditions carried out to grasp the driving pattern, instant values of fuel consumption and tailpipe emissions be gathered along with vehicle speed and location data.

It would be interesting to include the clustering step within the DC construction process. In that case, we recommend including the specific fuel consumption and emission indexes as criteria in the clustering process.

Being a stochastic method, the EBMT is repeatable but not reproducible. As described previously, aiming at reducing the problem of reproducibility, we propose to repeat the method a large number of times (~1000) and select as the representative DC the one with the lowest $ARDi$ value (Eq. 5.3).

5.4   The Markov Chain-Monte Carlo (MCMC) Method

The Markov Chain-Monte Carlo (MCMC) Method, also called the state-based or Markov chain approach, uses the Markov Chain theory, which assumes that the likelihood of a particular state (modal operation or vehicle kinematic chain) depends only upon the previous modal state.

This method assumes that the driving pattern is described by the speed-acceleration probability distribution (SAPD) described in Chapter 3. Thus, this method starts by discretizing the speed and acceleration ranges into n and m snippets, forming a state matrix (**SAPD**) of $n \times m$, as shown in Fig. 5.4. Then, the local driving pattern is described by the congruent speed acceleration probability distribution ($SAPD_{i,j}$). It is computed using the second-by-second data from each trip database, as described in Sect. 3.3 [27].

Next, using the same database of trips, a probability transition matrix (**T**) of $nm \times nm$ is built by computing the probability of the vehicle moving from state $SAPD_{i,j}$ to any state $SAPD_{i,j}$ of the state matrix (Fig. 5.5). Those probabilities are computed by counting the number of times vehicles moved from one state to another and then normalizing those frequencies to the total number of transitions.

Subsequently, the Monte Carlo (MC) technique is applied to assemble a candidate DC. It starts with the vehicle in idling mode, corresponding to the initial vehicle state (V_1). Next, for each iteration or vehicle state (V_k), the MC technique quasi-randomly selects the next vehicle state (V_{k+1}) based on the transition probability contained in **T**. It means

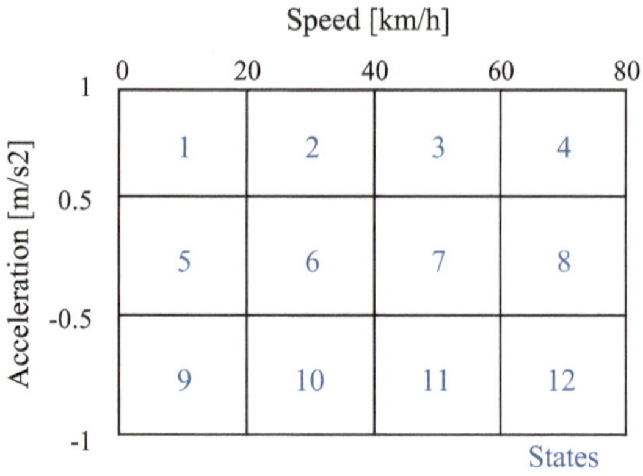

Fig. 5.4 Illustration of a state matrix or speed-acceleration probability distribution (SAPD) of 12 states

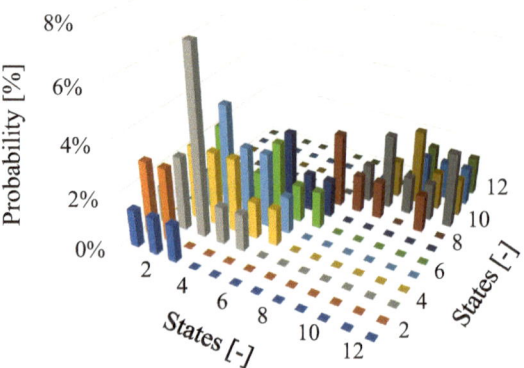

Fig. 5.5 Example of probability transition matrix (T) for states in Fig. 5.4

that a random number selects the next state and that the following potential states with the highest probability are the most likely to be chosen.

The set of consecutive states constitutes the states vector (V). Subsequently, this vector is decoded in terms of speed versus time producing a candidate DC [27, 30]. The process of selecting new states continues until the candidate DC has reached a pre-established time duration or traveled distance. Then, the representativeness of the candidate DC is evaluated through Eq. 5.2. However, the assessment criteria regarding CPs could also be used (Eq. 5.1). The algorithm restarts when the assessment criteria are not fulfilled. Otherwise, the candidate DC becomes the representative DC. Usually, a last step is carried out to smooth the speed versus time profile of the DC to reduce the artificial high accelerations that could be created in the transitions between states.

The MCMC method produces DC that ends with the vehicle at any speed. Like the MT method, the MCMC method is reproducible but not repeatable. Among the DCs that have used this approach is the LA01 cycle [30].

Recently, a relevant modification of this method has been reported. It replaces the state matrix **SAPD** with a state matrix of operational modes [27]. Each operational mode is defined by a kinematic sequence, which is obtained by splitting the trips according to pre-established criteria. They split the trips into segments of equal duration and then classified them according to their mean speed or similar speed acceleration. Given that the initial and final speeds are not zero, this alternative could present important problems, such as large and unrealistic artificially created accelerations when connecting two operational modes in the candidate DC.

5.5   The Trip-Based (TB) Method

The trip-based (TB) method is one of the oldest methods for constructing DC. In it, a whole trip from the collected trips is selected as the representative DC. It starts by describing the local driving pattern through any of the methods described in Sect. 2.7, being CPs the most frequent. Then, each trip is considered as a candidate DC.

Like the other methods, the representativeness of each candidate DC is evaluated by any of the methods described previously. Equation 5.1 is the preferred alternative where the average speed, mean positive acceleration, and percentage of idling time are the most recurrent CPs used. This method selects the candidate DC with the smallest average relative difference (Eq. 5.3) as the representative DC. This method is repeatable and reproducible and eliminates the problem of DC with unrealistic accelerations.

Some authors have expressed representativeness in terms of vehicle fuel consumption [2, 27]. They have argued that a DC that reproduces fuel consumption implicitly reproduces other CPs and emissions. Thus, the trip (or candidate DC) with the closest fuel consumption to the average value of fuel consumption of the entire trip sample is selected as the representative DC [40, 42]. When determining fuel consumption is impossible, a method (MWD-CP) was proposed to estimate the trip fuel consumption based on CPs.

The trip-based methods are appropriate when the sampled trips have similar duration, which is an unfrequent situation. Furthermore, it remains unresolved what to do when the trips are too long or too short. Section 5.6 will address the DC duration issue.

5.6   Artificial or Modal Driving Cycles

This method aims to construct a set of trapezes of speeds that represent the driving pattern of a given region or application. Each trapezoid of speeds includes four modes of vehicle operation: acceleration, cruise, deceleration, and idling (Fig. 5.6). The acceleration and cruise speed are selected quasi-randomly from the SAPD diagram using the Monte Carlo

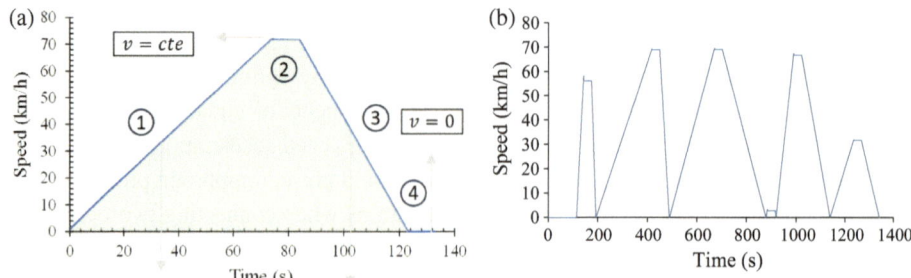

Fig. 5.6 Artificial or modal driving cycles. **a** Vehicle operational modes included in each trapezoid of speeds. **b** Illustration of an artificial DC that represents the driving patterns of long-distance freight transport in Ecuador

Technique. i.e., accelerations and speed with high probability in the SAPD are proportionally the most likely to be selected. Idling and cruise time are selected to equal the percentage of idling and cruise time observed in the driving pattern. The first approach is to use the same idling time in each trapezoid of speeds.

A collection of trapezoids of speed with a duration longer than a pre-established time forms a candidate DC. Then, its representativeness is evaluated using the means of a set of CPs like the MT method. However, in this case, the criteria of representativeness can be extended in the way that the candidate DC reproduces all known CPs, the SAPD and VSP distribution diagrams, and the energy consumption.

We suggest extending the criteria of representativeness further to include the effects of road grade variations in non-flat regions. Road grades in the DC can be included by obtaining a frequency distribution diagram of road grades similar to the VSP frequency distribution based on the data collected during the monitoring campaign carried out to grasp the driving pattern. Then, during the construction of the DC, besides selecting speed and acceleration, the Montecarlo technique could also be used to select a road grade for the acceleration, cruise, and deceleration modes. Thus, each trapezoid of speed is accompanied by three road grades, and the candidate DC is a collection of trapezoids of speed and road grades. i.e., a DC is a time series of speeds and road grades. The representativeness of the candidate DC includes reproducing road grade distribution frequency.

The implementation of such a DC on a chassis dynamometer requires the dynamometer to be programmable with a dynamic contra-torque to simulate the forces opposing the vehicle movement (i.e., aerodynamic, rolling, and road grade). Another alternative is to take the effect of road grade and acceleration and observe their combined effect on the engine torque demand. Then, replace the acceleration with an equivalent acceleration that demands the same torque to the engine.

5.7   DC Duration

Another parameter that impacts the representativeness of the DC and that has not been thoroughly analyzed is the duration of the DCs. When the DC is too short, it tends to misrepresent the driving patterns and increases the uncertainty in determining the vehicles' energy consumption and tailpipe emissions. Too long DCs demand excessive resources during the development of the tests where the DCs are used.

Aiming to define the duration of driving cycles, researchers have used their experience and knowledge of driving patterns in the region, mobility, and traffic surveys or have taken the duration of type approval cycles as a reference. It is suggested that a cycle generated by the MT method should last between 10 and 30 min [52]. Ho et al. recommended no less than 1000 s (16.6 min) [6].

Giraldo et al. [53] proposed a method for defining the appropriate duration of representative driving cycles. It is based on the observation that short DCs are less likely to fulfill representativeness requirements. Then, the appropriate DC duration is the shortest time when the DC construction method of interest tends to produce representative DC with the highest probability. Then, using the EBMT method, they obtained a large number (~500) of representative DCs starting from the same input database of trips. Then, they repeated the process for different DC time durations (5, 10, 15, 20, 25, 30, 45, 60 and 120 min). They observed the tendency and dispersion of the relative differences (RDi) of all known CPi and the specific fuel consumption and emissions indexes for each time duration. The tendency was associated with the average value of the ARD (Eq. 5.3), and the probability was related to the ARD's dispersion (interquartile range). As expected, they observed that increasing the duration of the DC, the $ARDj$ tends to 0, and their dispersion tends to reduce (Fig. 5.7). In this way, Giraldo et al. suggested that the duration of a driving cycle should correspond to that one that tends to produce $ARDj$ for CPs and EIs of less than 10% [53]. According to Fig. 5.7, the recommended duration of the DCs is about 20 min.

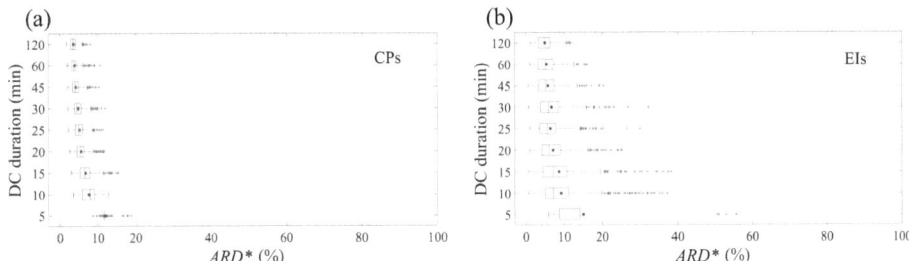

Fig. 5.7 Tendency (average value) and probability (interquartile range) of the average relative differences (ARD) obtained after 500 repetitions of obtaining representative DC by means of the EBMT method using the same input database of trips and varying DC duration. **a** Characteristic parameters and **b** emission indexes. *Source* [53]

5.8   Concluding Remarks

This chapter described the methodologies and their variations, most frequently used by researchers interested in constructing representative DCs. The micro-trips (MT) method was acknowledged as being the most recurrent. In it, the average speed, average positive acceleration, and percentage of idling time are the characteristic parameters most used as criteria of representativeness. There are many variations of this method. Nowadays, the energy-based micro-trips method is the most accepted because it was designed to represent vehicle energy consumption. The MT method does not have problems related to the easy reproduction of the resulting DCs during the development of tests conducted on the road and a chassis dynamometer. However, this method produces different DCs when reproduced with the same input data. i.e., it is a non-repeatable method. We suggested alternatives to mitigate this problem.

Among other methods to construct DC is the Monte Carlo-Markov Chain (MCMC) method, which, in addition to the repeatability issue, has problems with artificial and unrealistic accelerations.

We also described a method to construct artificial DCs, which is the easiest to reproduce when testing the vehicles. We suggested updating this method to ensure the full representativeness of the driving pattern. Furthermore, we showed its potential to address the inclusion of road grade effects within the construction method. It also can be transferable from region to region, reducing the cost of the monitoring campaign required to obtain the input data.

Finally, the effects of the DC duration on the level of representativeness of the driving pattern obtained with the resulting DCs were discussed. We concluded that 20 min seems to be the minimum time for ensuring that the DCs obtained truly represent driving patterns with a high probability.

References

1. Brady, J., O'Mahony, M.: Development of a driving cycle to evaluate the energy economy of electric vehicles in urban areas. Appl. Energy **177**, 165–178 (2016). https://doi.org/10.1016/j.apenergy.2016.05.094
2. Tong, H.Y., Hung, W.T.: A framework for developing driving cycles with on-road driving data. Transp. Rev. **30**, 589–615 (2010). https://doi.org/10.1080/01441640903286134
3. Galgamuwa, U., Perera, L., Bandara, S.: Developing a general methodology for driving cycle construction: comparison of various established driving cycles in the world to propose a general approach. J. Transp. Technol. **05**, 191–203 (2015). https://doi.org/10.4236/jtts.2015.54018
4. Huertas, J., Quirama, L., Giraldo, M., Díaz, J.: Comparison of three methods for constructing real driving cycles. Energies (Basel) **12**, 665 (2019). https://doi.org/10.3390/en12040665
5. Huertas, J.I., Quirama, L.F., Giraldo, M.D., Diaz, J.: Comparison of driving cycles obtained by the micro-trips, Markov-Chains and MWD-CP methods. Int. J. Sustain. Energy Plan. Manag. **22** (2019). https://doi.org/10.5278/ijsepm.2554

6. Ho, S.-H., Wong, Y.-D., Chang, V.W.-C.: Developing Singapore driving cycle for passenger cars to estimate fuel consumption and vehicular emissions. Atmos. Environ. **97**, 353–362 (2014). https://doi.org/10.1016/j.atmosenv.2014.08.042

7. Arun, N.H., Mahesh, S., Ramadurai, G., Shiva Nagendra, S.M.: Development of driving cycles for passenger cars and motorcycles in Chennai, India. Sustain. Cities Soc. **32**, 508–512 (2017). https://doi.org/10.1016/j.scs.2017.05.001

8. Lin, J., Niemeier, D.A.: An exploratory analysis comparing a stochastic driving cycle to California's regulatory cycle. Atmos. Environ. **36**, 5759–5770 (2002). https://doi.org/10.1016/S1352-2310(02)00695-7

9. Peng, Y., Zhuang, Y., Yang, Y.: A driving cycle construction methodology combining k-means clustering and Markov model for urban mixed roads. Proc. Inst. Mech. Eng. Part D: J. Autom. Eng. **234**, 714–724 (2020). https://doi.org/10.1177/0954407019848873

10. He, H., Sun, C., Zhang, X.: A method for identification of driving patterns in hybrid electric vehicles based on a LVQ neural network. Energies (Basel) **5**, 3363–3380 (2012). https://doi.org/10.3390/en5093363

11. Pouresmaeili, M.A., Aghayan, I., Taghizadeh, S.A.: Development of Mashhad driving cycle for passenger car to model vehicle exhaust emissions calibrated using on-board measurements. Sustain. Cities Soc. **36**, 12–20 (2018). https://doi.org/10.1016/j.scs.2017.09.034

12. Nyberg, P., Frisk, E., Nielsen, L.: Driving cycle equivalence and transformation. IEEE Trans. Veh. Technol. **66**, 1963–1974 (2017). https://doi.org/10.1109/TVT.2016.2582079

13. Gh, M.M., Naghizadeh, M.: Development of the Tehran car driving cycle. Int. J. Environ. Pollut. **30**, 106 (2007). https://doi.org/10.1504/IJEP.2007.014506

14. Yang, Y., Li, T., Hu, H., Zhang, T., Cai, X., Chen, S., Qiao, F.: Development and emissions performance analysis of local driving cycle for small-sized passenger cars in Nanjing, China. Atmos. Pollut. Res. **10**, 1514–1523 (2019). https://doi.org/10.1016/j.apr.2019.04.009

15. Yang, Y., Li, T., Zhang, T., Yu, Q.: Time dimension analysis: comparison of Nanjing local driving cycles in 2009 and 2017. Sustain. Cities Soc. **53**, 101949 (2020). https://doi.org/10.1016/j.scs.2019.101949

16. Xing, J., Xuefeng, H., Hui, Y., Yan, C., Huiping, Y.: Driving cycle recognition for hybrid electric vehicle. In: IEEE Transportation Electrification Conference and Expo, ITEC Asia-Pacific 2014 - Conference Proceedings, pp. 1–6 (2014). https://doi.org/10.1109/ITEC-AP.2014.6940693

17. Galgamuwa, U., Perera, L., Bandara, S.: Development of a driving cycle for Colombo, Sri Lanka: an economical approach for developing countries. J. Adv. Transp. **50**, 1520–1530 (2016). https://doi.org/10.1002/atr.1414

18. Nguyen, Y.-L.T., Nghiem, T.-D., Le, A.-T., Bui, N.-D.: Development of the typical driving cycle for buses in Hanoi, Vietnam. J. Air Waste Manag. Assoc. **69**, 423–437 (2019). https://doi.org/10.1080/10962247.2018.1543736

19. Zhao, X., Zhao, X., Yu, Q., Ye, Y., Yu, M.: Development of a representative urban driving cycle construction methodology for electric vehicles: a case study in Xi'an. Transp. Res. D Transp. Environ. **81**, 102279 (2020). https://doi.org/10.1016/j.trd.2020.102279

20. Ma, R., He, X., Zheng, Y., Zhou, B., Lu, S., Wu, Y.: Real-world driving cycles and energy consumption informed by large-sized vehicle trajectory data. J. Clean. Prod. **223**, 564–574 (2019). https://doi.org/10.1016/j.jclepro.2019.03.002

21. Gong, Q.; Midlam-Mohler, S.; Marano, V.; Rizzoni, G. An Iterative Markov Chain Approach for Generating Vehicle Driving Cycles. *SAE Int J Engines* **2011**, 4, 2011–01–0880, https://doi.org/10.4271/2011-01-0880.

22. Hongwen, H., Jinquan, G., Jiankun, P., Huachun, T., Chao, S.: Real-time global driving cycle construction and the application to economy driving pro system in plug-in hybrid electric vehicles. Energy **152**, 95–107 (2018). https://doi.org/10.1016/j.energy.2018.03.061

23. Zhang, J., Wang, Z., Liu, P., Zhang, Z., Li, X., Qu, C.: Driving cycles construction for electric vehicles considering road environment: a case study in Beijing. Appl. Energy **253**, 113514 (2019). https://doi.org/10.1016/j.apenergy.2019.113514
24. Esser, A., Zeller, M., Foulard, S., Rinderknecht, S.: Stochastic synthesis of representative and multidimensional driving cycles. SAE Int. J. Alternat. Powertrains **7** (2018), 2018–01–0095. https://doi.org/10.4271/2018-01-0095
25. Bishop, J.D.K., Axon, C.J., McCulloch, M.D.: A robust, data-driven methodology for real-world driving cycle development. Transp. Res. D Transp. Environ. **17**, 389–397 (2012). https://doi.org/10.1016/j.trd.2012.03.003
26. Wang, Z., Zhang, J., Liu, P., Qu, C., Li, X.: Driving cycle construction for electric vehicles based on Markov Chain and Monte Carlo method: a case study in Beijing. Energy Procedia **158**, 2494–2499 (2019). https://doi.org/10.1016/j.egypro.2019.01.389
27. Shi, S., Lin, N., Zhang, Y., Cheng, J., Huang, C., Liu, L., Lu, B.: Research on Markov property analysis of driving cycles and its application. Transp. Res. D Transp. Environ. **47**, 171–181 (2016). https://doi.org/10.1016/j.trd.2016.05.013
28. Gong, H., Zou, Y., Yang, Q., Fan, J., Sun, F., Goehlich, D.: Generation of a driving cycle for battery electric vehicles: a case study of Beijing. Energy **150**, 901–912 (2018). https://doi.org/10.1016/j.energy.2018.02.092
29. Yang, Y., Zhang, Q., Wang, Z., Chen, Z., Cai, X.: Markov Chain-based approach of the driving cycle development for electric vehicle application. Energy Procedia **152**, 502–507 (2018). https://doi.org/10.1016/j.egypro.2018.09.201
30. Li, Y., He, H., Peng, J.: An adaptive online prediction method with variable prediction horizon for future driving cycle of the vehicle. IEEE Access **6**, 33062–33075 (2018). https://doi.org/10.1109/ACCESS.2018.2840536
31. Hereijgers, K., Silvas, E., Hofman, T., Steinbuch, M.: Effects of using synthesized driving cycles on vehicle fuel consumption. IFAC-PapersOnLine **50**, 7505–7510 (2017). https://doi.org/10.1016/j.ifacol.2017.08.1183
32. Quirama, L.F., Giraldo, M., Huertas, J.I., Jaller, M.: Driving cycles that reproduce driving patterns, energy consumptions and tailpipe emissions. Transp. Res. D Transp. Environ. **82**, 102294 (2020). https://doi.org/10.1016/j.trd.2020.102294
33. Shen, P., Zhao, Z., Li, J., Zhan, X.: Development of a typical driving cycle for an intra-city hybrid electric bus with a fixed route. Transp. Res. D Transp. Environ. **59**, 346–360 (2018). https://doi.org/10.1016/j.trd.2018.01.032
34. Zou, S., Zhu, X., Xu, L., Zhu, H.: Construction of vehicle driving cycle in Fuzhou. J. Phys. Conf. Ser. **1453**, 012146 (2020). https://doi.org/10.1088/1742-6596/1453/1/012146
35. Wu, X., Hu, C., Du, J.: Development of a driving cycle for city bus in Harbin of China. Int. J. Electr. Hybrid Veh. **7**, 104 (2015). https://doi.org/10.1504/IJEHV.2015.071063
36. Gao, X., Zhang, B., Xiong, X., Dong, M., Li, H.: Construction and analysis of the Dalian driving cycle. Int. J. Control Autom. **8**, 363–368 (2015). https://doi.org/10.14257/ijca.2015.8.6.35
37. He, H., Guo, J., Zhou, N., Sun, C., Peng, J.: Freeway driving cycle construction based on real-time traffic information and global optimal energy management for plug-in hybrid electric vehicles. Energies (Basel) **10**, 1796 (2017). https://doi.org/10.3390/en10111796
38. Liu, B., Shi, Q., He, L., Qiu, D.: A study on the construction of Hefei urban driving cycle for passenger vehicle. IFAC-PapersOnLine **51**, 854–858 (2018). https://doi.org/10.1016/j.ifacol.2018.10.100
39. Huang, D., Xie, H., Ma, H., Sun, Q.: Driving cycle prediction model based on bus route features. Transp. Res. D Transp. Environ. **54**, 99–113 (2017). https://doi.org/10.1016/j.trd.2017.04.038
40. Huertas, J., Giraldo, M., Quirama, L., Díaz, J.: Driving cycles based on fuel consumption. Energies (Basel) **11**, 3064 (2018). https://doi.org/10.3390/en11113064

41. Saleh, W., Kumar, R., Kirby, H., Kumar, P.: Real world driving cycle for motorcycles in Edinburgh. Transp. Res. D Transp. Environ. **14**, 326–333 (2009). https://doi.org/10.1016/j.trd.2009.03.003

42. Huertas, J.I., Díaz, J., Cordero, D., Cedillo, K.: A new methodology to determine typical driving cycles for the design of vehicles power trains. Int. J. Interact. Des. Manuf. (IJIDeM) **12**, 319–326 (2018). https://doi.org/10.1007/s12008-017-0379-y

43. Yuhui, P., Yuan, Z., Huibao, Y.: Development of a representative driving cycle for urban buses based on the K-means cluster method. Cluster Comput **22**, 6871–6880 (2019). https://doi.org/10.1007/s10586-017-1673-y

44. Wang, J., Huang, J., Xie, H., Tian, G.: Driving pattern prediction model for hybrid electric buses based on real-world driving data. In: Proceedings of the 28th International Electric Vehicle Symposium and Exhibition (2015)

45. Zhao, X., Yu, Q., Ma, J., Wu, Y., Yu, M., Ye, Y.: Development of a representative EV urban driving cycle based on a K-means and SVM hybrid clustering algorithm. J. Adv. Transp. **2018**, 1–18 (2018). https://doi.org/10.1155/2018/1890753

46. Pan, C., Gu, X., Chen, L., Chen, L., Yi, F.: Driving cycle construction and combined driving cycle prediction for fuzzy energy management of electric vehicles. Int. J. Energy Res. **45**, 17094–17108 (2021). https://doi.org/10.1002/er.5320

47. Hung, W.T., Tong, H.Y., Lee, C.P., Ha, K., Pao, L.Y.: Development of a practical driving cycle construction methodology: a case Study in Hong Kong. Transp Res D Transp Environ **12**, 115–128 (2007). https://doi.org/10.1016/j.trd.2007.01.002

48. Tamsanya, S., Chungpaibulpatana, S., Limmeechokchai, B.: Development of a driving cycle for the measurement of fuel consumption and exhaust emissions of automobiles in Bangkok during peak periods. Int. J. Automot. Technol. **10**, 251–264 (2009). https://doi.org/10.1007/s12239-009-0030-4

49. Nouri, P., Morency, C.: Evaluating microtrip definitions for developing driving cycles. Transp. Res. Record: J. Transp. Res. Board **2627**, 86–92 (2017). https://doi.org/10.3141/2627-10

50. Xie, H., Tian, G., Chen, H., Wang, J., Huang, Y.: A distribution density-based methodology for driving data cluster analysis: a case study for an extended-range electric city bus. Pattern Recogn. **73**, 131–143 (2018). https://doi.org/10.1016/j.patcog.2017.08.006

51. Huertas, J.I., Coello, G.A.Á.: Accuracy and precision of the drag and rolling resistance coefficients obtained by on road coast down tests. In: Proceedings of the International Conference on Industrial Engineering and Operations Management, pp. 575–582 (2017)

52. Amirjamshidi, G., Roorda, M.J.: Development of simulated driving cycles for light, medium, and heavy duty trucks: case of the Toronto waterfront area. Transp. Res. D Transp. Environ. **34**, 255–266 (2015). https://doi.org/10.1016/j.trd.2014.11.010

53. Giraldo, M., Quirama, L.F., Huertas, J.I., Tibaquirá, J.E.: The effect of driving cycle duration on its representativeness. World Electr. Veh. J. **12** (2021). https://doi.org/10.3390/wevj12040212

Driving Cycle Validation

6

Michael Giraldo, José I. Huertas, and Nicolas Giraldo Peralta

Abstract

After obtaining a DC, most authors compare it against other DCs, particularly type-approval DCs, and highlight the observed differences. However, few published works take actions to (i) ensure that the computational algorithm implemented is free of mistakes and (ii) verify that the obtained DC accurately represents the driving pattern of the region and replicates real-world vehicle fuel consumption and emissions when tested on a chassis dynamometer. This chapter outlines methodologies for evaluating the implemented computational algorithm, assessing the DC's ability to replicate energy consumption and emissions, and considerations for using a DC in specific applications.

M. Giraldo
Industry, Materials and Energy Area, School of Applied Sciences and Engineering, Universidad EAFIT, 050030 Medellín, Colombia
e-mail: mgiral36@eafit.edu.co

J. I. Huertas (✉)
Sustainable Energy Research Group, Tecnologico de Monterrey, Monterrey 64849, México
e-mail: jhuertas@tec.mx

N. Giraldo Peralta
School of Mechanical, Electronic, and Biomedical Engineering, Universidad Antonio Nariño, 111511 Bogotá, Colombia
e-mail: nicolas.giraldo@uan.edu.co

© The Author(s), under exclusive license to Springer Nature Switzerland AG 2025
J. I. Huertas (ed.), *Fundamentals of Driving Patterns and Driving Cycles*, Synthesis Lectures on Mechanical Engineering, https://doi.org/10.1007/978-3-031-76863-7_6

6.1   Tests to Verify the Correct Implementation of the DC Construction Method

We have not seen any work where authors verify the correct implementation of the selected construction method to obtain DCs besides in [1]. It could be argued that as long as the obtained DC complies with the desired values for the metrics of representativity selected, the obtained DC is correct. However, errors could exist in the code implemented to calculate the metric of representativeness. The reproducibility, but not repeatability, of the methods most commonly used to construct DCs makes identifying potential implementation errors challenging.

A common alternative to verify the correct implementation of any computational code is to compare the results obtained by the code when applied to a case where the solution is well known. However, there are no publicly available trip databases that could be used as input data for DC construction methods with a well-accepted output DC. Furthermore, since some of the DC construction methods are not repeatable, and therefore, multiple DCs can be accepted as representative DCs, this alternative has not been established yet. We suggest exploring this possibility further.

In the same line of thinking, Huertas et al. [1] suggested constructing an artificial but simple micro-trips database to serve as input data for evaluating the DC construction method. For example, the database could consist of many (~500) micro-trips, where each micro-trip has the same trapezoidal shape as shown in Fig. 6.1. This setup means that everybody in the city travels from place to place, following a connection of trapezoidal micro-trips. i.e., the DC is known and equal to the one represented by the trapezoidal micro-trip. Therefore, independently of the DC construction method, the resulting DC should be a collection of the same trapezoidal micro-trips.

The level of similitude between the trapezoidal micro-trip and the obtained DC indicates that the method, or its implementation, can capture the known driving pattern. Equations 6.1–6.5, which will be explained later in this chapter, could be used for this purpose. Huertas et al. [1] built the two artificial trips shown in Fig. 6.1a, b. The first trip consists of a single micro-trip with a starting idling time of 10 s, followed by a constant acceleration (0.28 m/s^2), a cruise speed of 80 km/h during 120 s, and a continuous desacceleration (-0.28 m/s^2). The second testing trip consists of 3 micro-trips, with different acceleration ramps and cruise speeds.

For example, Fig. 6.1b shows the result reported by the MCMC method using the database made of a single artificial micro-trip shown in Fig. 6.1a. Similarly, Fig. 6.1d. shows the DC obtained by the EBMT method using as input a database made of the trips shown in Fig. 6.1c. In both cases, the resulting DC is similar to the known result. It is highlighted that in DCs, the order in which the different micro-trips show up is irrelevant. DCs with the same collection of micro-trips but in different sequences lead to the same energy consumption and tailpipe emissions.

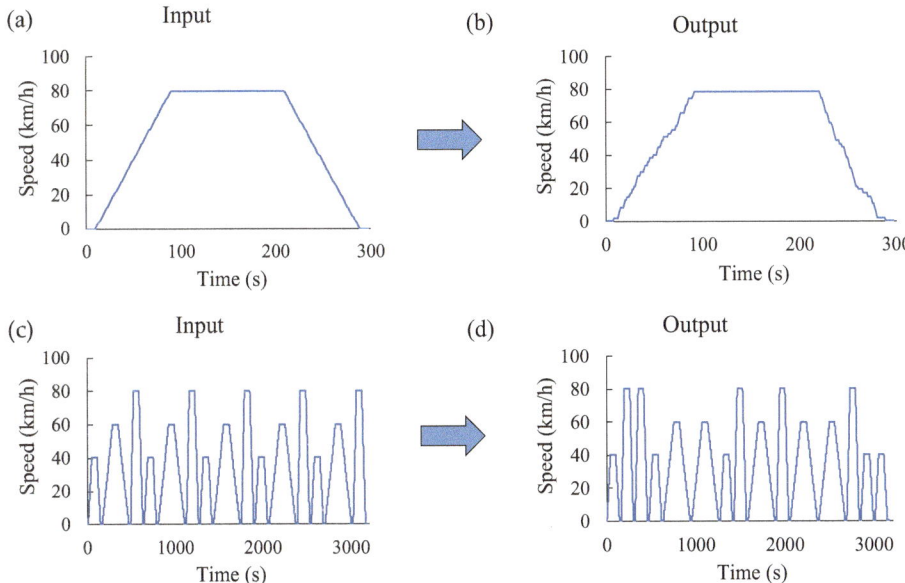

Fig. 6.1 Test used to verify the correct implementation of any computational algorithm to construct DCs. **a** Driving pattern consisting of **a** a single trapezoidal micro-trip and **c** a collection of three micro-trips. **b** Result obtained by an MCMC method using a database of many single trapezoidal micro-trip as inputs. **d** Result of the FBMT method when a database consisting of many triads of trapezoidal micro-trips is used as input. *Source* [1]

We suggest this test to verify the algorithm's correct implementation and the method's capacity to produce DCs that truly represent any driving patterns.

6.2   Comparison Between Methods to Construct DCs

This section describes a way to evaluate how well the DCs produced by any DC construction method represent the local driving pattern under study. We recall that driving patterns describe how people drive in a given region and that several alternatives exist to describe those driving patterns. The most frequent are (i) a collection of CPs, (ii) an SAPD, (iii) a VSP frequency distribution diagram, and (iv) DCs.

The correct representation of the local driving pattern through a DC depends mainly on two factors: (i) the quantity and quality of the vehicles' operation data used to describe the driving pattern, and (ii) the DC construction method [1–3]. The first factor ensures the quantity and quality of the data that properly captures the local driving pattern. This factor was discussed in Chap. 3. We now turn the discussion into the capacity of a given DC construction method to produce DCs that truly represent local driving patterns.

We have underlined that the main application of DCs is the characterization of the energy and environmental performance of the vehicles. Therefore, we propose to evaluate the performance of the construction method by the ability of the DCs produced by such a method to reproduce the energy consumption and emission factors observed in the everyday operation of the vehicles in the region under study. Furthermore, under the uncertainty of which alternative best describes the driving pattern, we can extend this proposal to the ability of the resulting DCs to reproduce all known CPs, the SAPD, and the VSP frequency distribution. Studies show that including all criteria increases the ability of the methods to reproduce driving patterns [4–6].

Figure 6.2 illustrates the methodology suggested to compare different DC construction methods. It starts with a common database gathered from a monitoring campaign of the operation of a vehicle fleet in a given region. It contains a collection of trips, and each trip consists of data at 1 Hz of location (longitude, latitude, and altitude), vehicle speed, fuel consumption, CO, and NOx emission rates. The driving pattern is described by a subset of the CPs listed in Table 3.1, the SAPD, and the VSP frequency distribution. The average values for SFC, EI_{CO}, and EI_{NOx} are also obtained.

Then, the DC construction method under evaluation is used with the common database as input. The resulting DC is used to obtain the CPs*, SAPD*, frequency distribution of VSPs*, SFC*, and EI*s. We recall that * refers to those of the DC while the absence of * to the driving pattern. Finally, the method's performance is evaluated by the relative differences between the variable observed to describe the driving pattern and the variable observed to describe the driving cycle. In Eqs. 6.1–6.3, the index i indicates any of the CPs listed in Table 3.1 while the index j indicates CO_2, CO, or NOx.

$$RD_{CPi} = \frac{|Cp_i - Cp_i^*|}{Cp_i} \tag{6.1}$$

$$RD_{SFC} = \frac{|SFC - SFC^*|}{SFC} \tag{6.2}$$

$$RD_{EIj} = \frac{|EI_j - EI_j^*|}{EI_j} \tag{6.3}$$

$$RD_{VSP} = 1 - \frac{|VSP \cdot VSP^*|}{\|VSP\| \, \|VSP^*\|} \tag{6.4}$$

$$RD_{SAPD} = 1 - \frac{|SAPD \cdot SAPD^*|}{\|SAPD\| \, \|SAPD^*\|} \tag{6.5}$$

We suggest using the cosine similarity concept to evaluate the relative difference in the VSP frequency distribution of the pattern and the DC (RD_{VSP}, Eq. 6.4). It measures the similarity between two vectors by means of the dot product between them. It defines the cosine angle between two vectors in the two-dimensional space. In this case, the frequency

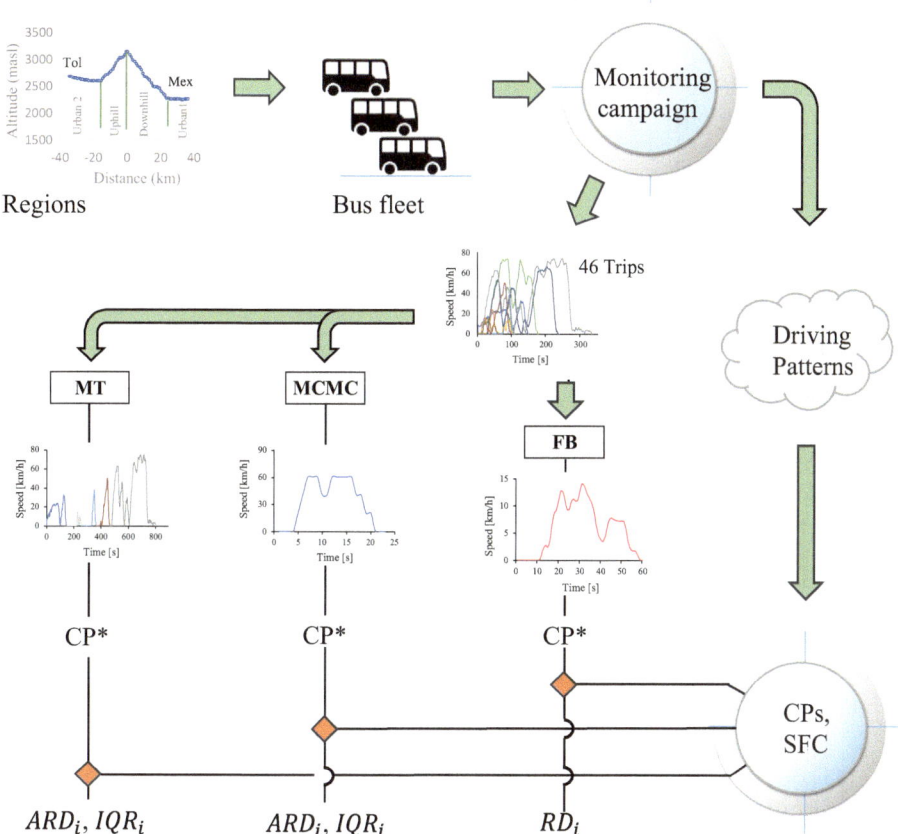

Fig. 6.2 Methodology suggested to compare DCs construction methods [1]

distribution of VSPs is described by vector **VSP**, while the VSP frequency distribution of the DC is described by vector **VSP***. Equation 6.4 | | indicates the absolute value, while ‖ ‖ indicates the magnitude of the vector. RD_{VSP} varies between 0 and 1, where 0 indicates perfect similarity and 1 no similarity at all.

Although the cosine similarity technique is used to compare the similarity between two vectors in the two-dimensional space, it has also been extended to the n-space, where n is the dimensions of the vector [7]. Thus, we used this concept to evaluate the similarity between the SAPD of the driving pattern and that of the DC, where again, **SAPD** is the matrix for the pattern, and **SAPD*** is the matrix for the DC.

Given that the most used DC construction methods are stochastic, and therefore, they are not repeatable (they do not produce the same result every time it is run with the same input data), it is recommended to run the DC construction method multiple (~500) iterations always using the same common database as input and report the tendencies of

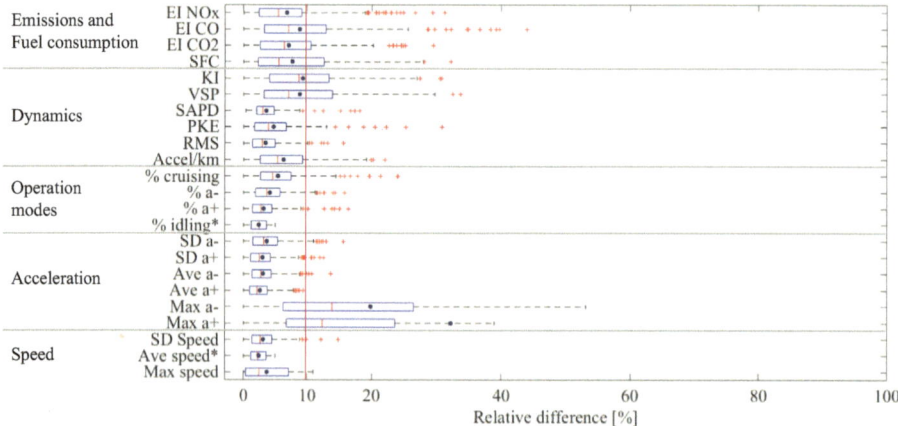

Fig. 6.3 Illustrative results of the distribution of relative differences for CPs, SAPD, VSP, emissions, and fuel consumption after 500 iterations using the fuel-based micro trip method (FBMT) with a database of trips made by 16 buses traveling between Toluca and Mexico City [1]

the relative differences. The tendencies can be presented as shown in Fig. 6.3 in terms of boxplots. This type of graph presents the distribution of the results in terms of the interquartile range (blue box), median (vertical red line in the boxes), mean (blue dots), and outliers (red + signs).

Values for the relative differences in Eqs. 6.1–6.3 can vary between zero and infinity (0–1 in Eqs. 6.4 and 6.5). Values close to zero are desirable. Values smaller than 10% for all variables are typically used as an acceptable threshold value for methods that produce representative DCs.

On the other hand, the size of blue boxes (interquartile range) is associated with the method's repeatability. Again, the interquartile range can vary between zero and infinity. The smaller the interquartile range, the better the repeatability of the method.

We suggest the use of the relative differences (Eqs. 6.1–6.5) to evaluate the performance of the method used to construct representative DCs. However, other studies have proposed different metrics. For the readers' reference, we list some of them in Table 6.1.

6.3   Considerations When Deploying a DC

As mentioned multiple times in this book, DCs are primarily used to evaluate the energy and environmental performance of ground vehicles. There are two test protocols to perform such evaluations. (i) On a chassis dynamometer, and (ii) On the road [21]. In both cases, a technician follows the DC by observing on a screen the instant speed at which the vehicle should be running. The technician compares it with the vehicle's speed and

Table 6.1 Some metrics reported in the literature to evaluate the performance of DC construction methods

Year	Region	DC construction method	Metric to evaluate differences	Reference
2023	Prishtine, Kosovo	MT	Coefficient of variation of speed	[8]
2023	Hong Kong, China	MT	Average absolute percentage error (AAPE)	[9, 10]
2023	Fuzhou, China	MT	Relative error	[11, 12]
2022	Tehran, Iran	MT	Relative positive acceleration (RPA)	[13, 14]
2022	Shenyang, China	MCMC	Mean absolute error (MAE)	[15]
2021	Islamabad, Pakistan	MCMC	Mean absolute percentage error (MAPE)	[16, 17]
2021	Greater Cairo, Egypt	MT	Percentage deviation (PD)	[18]
2021	Xi'an, China	MT	Absolute relative error (ARE) Mean absolute relative error (MARE)	[14]
2020	Xi'an China	MCMC	Root mean square error (RMSE)	[19, 20]

decides the acceleration required to minimize differences by pressing or depressing the accelerator pedal. This labor requires specialized skills. It is a risky activity where errors could cause even mortal fatalities. Therefore, these tests are stressful, time-consuming, and subject to several repetitions, and consequently, they are costly (Fig. 6.4).

For the same reasons, these tests are performed on a few samples of the same vehicle technology, and researchers rely on the fact that vehicles of the same technology will have the same performance. Therefore, researchers usually perform these tests with vehicles of different technologies. Since every technology has its own speed response to the pedal accelerator, replacing the technician with a robot is challenging.

Next, we will describe some considerations to consider when deploying a DC.

6.3.1 The Use of DCs When Performing Tests on a Chassis Dynamometer

As explained in Sect. 3.6, the evaluation of the energy and environmental performance of light-duty vehicles and motorcycles is usually performed on a chassis dynamometer

Fig. 6.4 Illustration of the process to follow a DC when testing road vehicles' energy and environ-mental performance

reproducing a DC. However, this practice is uncommon in the case of heavy-duty vehicles due to several practical considerations:

- Excess loads: Difficulties associated with the vehicle weight distribution on the rollers.
- The presence of several wheels on the same axle and the presence of double axes.
- It is not mandatory. The regulations do not demand these types of tests for the case of heavy-duty vehicles because the results of a single test do not represent the per-formance of a technology. Heavy-duty vehicles are often customized for specific applications. Currently, few truck manufacturers sell rolling chassis with or without a cabin, and multiple local companies customize the chassis for a particular use or specific application. For example, the same rolling chassis can be used as a bus for the transport of people or a truck for the transport of goods. Therefore, every heavy-duty vehicle needs to be considered an independent technology whose energy and environmental performance need to be evaluated separately, which is unpractical.

Then, instead of evaluating the performance of heavy-duty vehicles on a chassis dynamometer, manufacturers assess the engine's performance on an engine test bench. Few heavy-duty vehicles are evaluated on a chassis dynamometer. In this case, the inter-est is in the performance of the rolling chassis or the tractive unit, which, as described previously, is not representative of the vehicles running in a city.

Testing the performance of the vehicles following a DC on a chassis dynamometer under laboratory conditions offers the advantage that the effects of external and human factors are under control and, therefore, results are repeatable. Weather or traffic condi-tions do not affect results because tests are performed indoors under simulated working conditions. The use of the DC isolates the human influencing factor. Therefore, tests are reproducible and repeatable. Furthermore, these tests are appropriate when the effects of external variables need to be considered, such as cold-start or ambient temperature. These conditions can only be achieved in the confined environment of a test cell [22]. These characteristics are essential for comparative and regulatory purposes [23].

However, these tests on chassis dynamometer have several significant issues:

- In practice, the energy consumption and tailpipe emissions values obtained when testing vehicles on a chassis dynamometer differ from those observed during everyday vehicle operation. Despite the effort to obtain representative DCs that reproduce energy consumption and tailpipe emissions, vehicles behave differently on the streets than on a chassis dynamometer. There is still a gap in knowledge on how to simulate the real operating conditions on a chassis dynamometer.
- They do not include the effects of the road grade variable. It is assumed that the DC was obtained for a flat region.
- They do not consider potential variations in the engine's performance due to variations in altitude. Testing protocols, especially for regulatory purposes, require that the tests be conducted at sea level.
- The road resistance exerted on the vehicle on the chassis dynamometer is simulated using data derived from coast-down tests. In real life, road resistance varies on each vehicle wheel, depending mostly on the vehicle's acceleration conditions. Thus, the road resistance on a chassis dynamometer cannot precisely reflect the resistance on real-world roads.
- Dynamometer calibration is performed under static or steady-state conditions. The calibration of the chassis dynamometer must consider bench losses and rolling resistance, which are difficult to handle under dynamic experimental conditions.

6.3.2 The Use of DCs When Performing On-Road Tests

An alternative to overcome the issues of evaluating the vehicle's performance by reproducing a DC on chassis dynamometers is to carry on such tests on the road under controlled (test track) or everyday conditions (streets). Using tracks specially designed to test vehicles is expensive, but they offer infrastructure to reduce the risk of accidents and mitigate the impact of potential ones. Every day, the number of studies conducted directly on the streets is growing. In these studies, researchers take special measures to reduce the risk of accidents. This alternative is less expensive than previous alternatives. It is the most realistic but the riskiest, as well.

Since experimenters do not have control over whether conditions results exhibit more variability compared to the results obtained under laboratory conditions. This drawback can be weakened by increasing the number of repetitions of the same test under different weather conditions and letting statistics average those external effects.

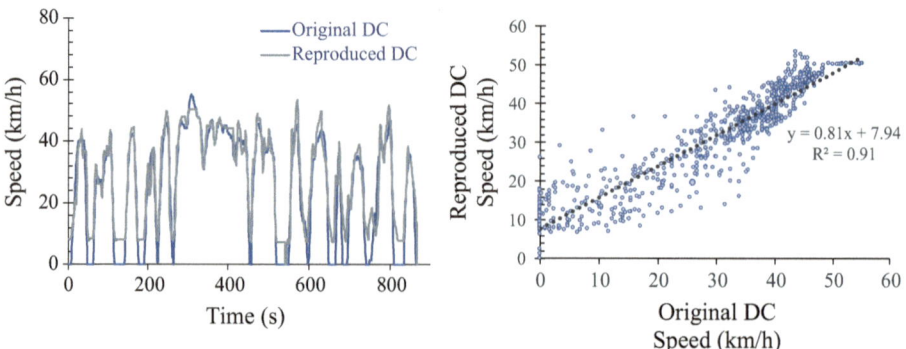

Fig. 6.5 Illustrative results of an evaluation of the degree of reproducibility of a DC

6.3.3 Verifying the Reproduction of a DC

Despite the infrastructure used to ease the reproduction of DC during vehicle tests and
the technician's training in this task, there would always be errors in the reproduction of
the selected DC. Acknowledging that the energy and environmental performance of the
vehicles depend directly on the reproduction of the DC, it is crucial to ensure the correct
reproduction of the DC. Therefore, it is relevant to define metrics to evaluate the degree
under which a DC was reproduced.

Agudelo et al. [24] suggested using the coefficient of determination (R^2) obtained in
a correlation analysis between the speed specified by the DC and the actual speed of
the vehicle during the duration of the tests. Figure 6.5 illustrates this analysis. From the
experience of replicating DC on chassis dynamometers and the road, they determined
that R^2 greater than 0.90 can be used as a threshold value for a successful or acceptable
reproduction of a given DC.

6.4 Concluding Remarks

In this chapter, we described the steps recommended to take after obtaining a DC by any
method with the purpose of demonstrating that the computational algorithm implemented
is free of mistakes and verify that the DC obtained actually represents the driving pattern
of the region and reproduces fuel consumption and emissions when vehicles follow the
obtained DC on a chassis dynamometer.

The essence of the validation process is to use an artificial or real database of trips as
input to the DC construction method under evaluation and verify it produces, systematically, DCs with CPs*, SAPD*, frequency distribution of VSPs*, SFC*, and EI*s like the

actual CPs, SAPD, frequency distribution of VSPs, SFC, and EIs of the driving pattern. Note that * is used for the values obtained with the DC.

References

1. Huertas, J., Quirama, L., Giraldo, M., Díaz, J.: Comparison of three methods for constructing real driving cycles. Energies (Basel) **12**, 665 (2019). https://doi.org/10.3390/en12040665
2. Huertas, J., Giraldo, M., Quirama, L., Díaz, J.: Driving cycles based on fuel consumption. Energies (Basel) **11**, 3064 (2018). https://doi.org/10.3390/en11113064
3. Huertas, J.I., Quirama, L.F., Giraldo, M.D., Diaz, J.: Comparison of driving cycles obtained by the micro-trips, Markov-chains and MWD-CP methods. Int. J. Sustain. Energy Plan. Manag. **22** (2019). https://doi.org/10.5278/ijsepm.2554
4. Quirama, L.F., Giraldo, M., Huertas, J.I., Jaller, M.: Driving cycles that reproduce driving patterns, energy consumptions and tailpipe emissions. Transp. Res. D Transp. Environ. **82**, 102294 (2020). https://doi.org/10.1016/j.trd.2020.102294
5. Giraldo, M., Quirama, L.F., Huertas, J.I., Tibaquirá, J.E.: The effect of driving cycle duration on its representativeness. World Electr. Veh. J. **12** (2021). https://doi.org/10.3390/wevj12040212
6. Quirama, L.F., Giraldo, M., Huertas, J.I., Tibaquirá, J.E., Cordero-Moreno, D.: Main characteristic parameters to describe driving patterns and construct driving cycles. Transp. Res. D Transp. Environ. **97** (2021). https://doi.org/10.1016/j.trd.2021.102959
7. Meyer, C.D., Stewart, I.: Matrix Analysis and Applied Linear Algebra, 2nd edn. Society for Industrial and Applied Mathematics, Philadelphia, PA (2023). ISBN: 978-1-61197-743-1
8. Salihu, F., Demir, Y.K., Demir, H.G.: Effect of road slope on driving cycle parameters of urban roads. Transp. Res. D Transp. Environ. **118**, 103676 (2023). https://doi.org/10.1016/j.trd.2023.103676
9. Tong, H.Y., Ng, K.W.: Developing electric bus driving cycles with significant road gradient changes: a case study in Hong Kong. Sustain. Cities Soc. **98**, 104819 (2023). https://doi.org/10.1016/j.scs.2023.104819
10. Tong, H.Y., Ng, K.: Development of bus driving cycles using a cost effective data collection approach. Sustain. Cities Soc. **69**, 102854 (2021). https://doi.org/10.1016/j.scs.2021.102854
11. Lin, J., Liu, B., Zhang, L.: Autoencoder-based optimization method for driving cycle construction: a case study in Fuzhou, China. J. Ambient. Intell. Humaniz. Comput. **14**, 12635–12650 (2023). https://doi.org/10.1007/s12652-022-04317-7
12. Zhang, L., Huang, Z., Yu, F., Liao, S., Luo, H., Zhong, Z., Zhu, M., Li, Z., Cui, X., Yan, M., et al.: Road type-based driving cycle development and application to estimate vehicle emissions for passenger cars in Guangzhou. Atmos. Pollut. Res. **12**, 101138 (2021). https://doi.org/10.1016/j.apr.2021.101138
13. Mafi, S., Kakaee, A., Mashadi, B., Moosavian, A., Abdolmaleki, S., Rezaei, M.: Developing local driving cycle for accurate vehicular CO_2 monitoring: a case study of Tehran. J. Clean. Prod. **336**, 130176 (2022). https://doi.org/10.1016/j.jclepro.2021.130176
14. Wang, L., Ma, J., Zhao, X., Li, X.: Development of a typical urban driving cycle for battery electric vehicles based on kernel principal component analysis and random forest. IEEE Access **9**, 15053–15065 (2021). https://doi.org/10.1109/ACCESS.2021.3052820
15. Chen, Z., Fang, Z., Zhang, Q., Zhou, N., Yu, Q.: Constructing the real-world driving cycle for electric vehicle applications: a comparative study. Trans. Inst. Meas. Control (2022). https://doi.org/10.1177/01423312221094384

16. Qin, X., Yu, K., Li, H., Dai, F., Liu, H., Yang, H., Ye, J., Zhu, H.: Development of a one-day driving cycle for electric ride-hailing vehicles. Transp. Res. D Transp. Environ. **89**, 102597 (2020). https://doi.org/10.1016/j.trd.2020.102597
17. Bhatti, A.H.U., Kazmi, S.A.A., Tariq, A., Ali, G.: Development and analysis of electric vehicle driving cycle for hilly urban areas. Transp. Res. D Transp. Environ. **99**, 103025 (2021). https://doi.org/10.1016/j.trd.2021.103025
18. Huzayyin, O.A., Salem, H., Hassan, M.A.: A representative urban driving cycle for passenger vehicles to estimate fuel consumption and emission rates under real-world driving conditions. Urban Clim. **36**, 100810 (2021). https://doi.org/10.1016/j.uclim.2021.100810
19. Zhao, X., Zhao, X., Yu, Q., Ye, Y., Yu, M.: Development of a representative urban driving cycle construction methodology for electric vehicles: a case study in Xi'an. Transp. Res. D Transp. Environ. **81**, 102279 (2020). https://doi.org/10.1016/j.trd.2020.102279
20. Chauhan, B.P., Joshi, G.J., Purnima, P.: Candidate driving cycle construction for emission estimation. In: Mathew, T.V., Joshi, G.J., Velaga, N.R., Arkatkar, S. (eds.) Proceedings of the Transportation Research. Springer Singapore, Singapore, pp. 85–97 (2020)
21. Li, T., Chen, X., Yan, Z.: Comparison of fine particles emissions of light-duty gasoline vehicles from chassis dynamometer tests and on-road measurements. Atmos. Environ. **68**, 82–91 (2013). https://doi.org/10.1016/j.atmosenv.2012.11.031
22. Franco, V., Kousoulidou, M., Muntean, M., Ntziachristos, L., Hausberger, S., Dilara, P.: Road vehicle emission factors development: a review. Atmos. Environ. **70**, 84–97 (2013). https://doi.org/10.1016/j.atmosenv.2013.01.006
23. Chen, L., Wang, Z., Liu, S., Qu, L.: Using a chassis dynamometer to determine the influencing factors for the emissions of Euro VI vehicles. Transp. Res. D Transp. Environ. **65**, 564–573 (2018). https://doi.org/10.1016/j.trd.2018.09.022
24. Agudelo-Santamaría, J., Agudelo-Santamaría, A.F., López-García, A.F., Giraldo-Peralta, N.: Factores de Emisión de Los Combustibles Colombianos (FECOC+) Fase 2.2: Determinación de Los Factores de Emisión de Vehículos Pesados de Carga (Camiones y Tractocamiones) y de Pasajeros (Buses) a La Altitud de Bogotá y Barranquilla (2023)

Comparing Driving Patterns

José I. Huertas⦿ and Rogelio Escamilla Serrano⦿

Abstract

Through a practical case study, this chapter shows that CPs can be associated with traffic conditions (external factors), SAPDs with driving styles (human factors), and VSP with fuel consumption (technological factors). Furthermore, we show how the use of dimensionless numbers to describe driving patterns enables the development of fuel consumption reduction strategies tailored to specific regions and situations.

7.1 Introduction

So far, we have discussed how to describe driving patterns, construct DCs, and verify that the DCs constructed effectively describe how people drive in a given region. After completing these objectives, a key question remains: How can we compare driving patterns to identify new strategies for reducing vehicle fuel consumption?

There are two main groups interested in this question. First, local governments are interested in the external factors influencing driving patterns and in identifying strategies related to the country's infrastructure to reduce the fuel consumption of the transport sector while increasing its productivity to reduce the national greenhouse gas emissions. The second group is made up of fleet managers who are interested in reducing the operational

J. I. Huertas (✉) · R. Escamilla Serrano
Sustainable Energy Research Group, Tecnologico de Monterrey, Monterrey 64849, México
e-mail: jhuertas@tec.mx

R. Escamilla Serrano
e-mail: rogelio.escam@outlook.com

© The Author(s), under exclusive license to Springer Nature Switzerland AG 2025
J. I. Huertas (ed.), *Fundamentals of Driving Patterns and Driving Cycles*, Synthesis Lectures on Mechanical Engineering, https://doi.org/10.1007/978-3-031-76863-7_7

costs of the transport service. They are interested in reducing fuel consumption by focusing on the human factors influencing driving patterns. i.e., they are interested in comparing driving styles, which is the core concept of any eco-driving program's objective.

Thus, the question of how to compare driving patterns for obtaining useful information includes the development of methodologies to isolate external, technological, and human factors from each other.

Aiming to address partially these questions, this chapter compares the driving patterns observed in different regions. We will use this information to isolate the influencing factors. It is highlighted that the objective of this chapter is not to compare DCs but to compare driving patterns.

7.2   Case of Study

Aiming to illustrate the methods to compare driving patterns with the purpose of obtaining useful information toward the reduction of fuel consumption and tailpipe emissions, we will use a case study where the effects of the technology factor can be isolated from the driving patterns. Thus, we selected a single technology and observed the driving patterns of drivers that use this technology in different regions. The technology selected was a heavy-duty truck of the International Prostar+ brand, equipped with a 450-horsepower diesel engine and an 18-speed transmission. Table 7.1 shows the technical characteristics of the technology selected.

The regions selected were Mexico, Ecuador, and Colombia. To facilitate the data analysis, we observed the operation of the same technology serving a single route in each region. Figure 7.1 shows the routes selected along with a plot of the road altitude variation.

The study regions were selected for their topographic diversity. In the case of Mexico, urban routes characterized by predominantly flat terrain were chosen. They are 1600 m above sea level (masl) and have an average temperature of 18–22 °C. For Ecuador, the selected routes exhibit a significant variation in elevation between the start and end of each route, situated at an average altitude of 1,235 masl, with an average temperature range of 15–27 °C. Similarly, in the Colombian region, routes with steeper slopes were analyzed, reaching altitudes of up to 3,500 masl, with average temperatures ranging between 7 and 19 °C.

Telemetry systems monitored the vehicles in the same way as described in previous chapters. The monitoring periods varied between countries according to the possibilities agreed upon with the telemetry provider but were long enough to ensure the representativeness of the data obtained. Table 7.2 includes the number of trips monitored and the periods of observation.

Table 7.1 Technical characteristics of the technology selected to compare the driving patterns in different regions in LATAM

Vehicle	Characteristic	Description	
International Prostar+ 122 6 × 4	Engine	CUMMIS ISX 450 450HP/18000 1650 lb-ft/ 1200 6 cylinders, 14.9 L, Turbocharged, EPA98, Electronic injection Diesel	
	Transmission	Fuller RTLO-16918B, 18 speeds with double overdrive 6 × 4 traction	
		Forward gears: 14.4, 12.29, 8.56, 7.3, 6.05, 5.16, 4.38, 3.74, 3.2, 2.73, 2.29, 1.95, 1.62, 1.38, 1.17, 1.00, 0.86, 0.73	
		Differential: RT-46-160P Corona ratio 4.3	
		RPM range	1200
	Wheel	11R22.5	
	Weight	Curb weight	8,160 kg
		GVWR	27,400 kg
	Frontal area	10.00 m^2	
	Drag coefficient	0.510	

7.3   Comparison in Terms of CPs, SAPDs, and VSPs

As described before, driving patterns can be described using CPs, SAPDs, VPSs, and DCs. We cannot use DCs to compare driving patterns because, as explained before, many DCs can represent a single driving pattern. Therefore, using DCs to compare driving patterns will introduce undesirable noise into the analysis.

Comparing Driving Patterns by Means of CPs
The lack of a clear association of CPs with traffic in a city has limited the use of CPs for decision-making processes related to strategies to improve mobility. This issue can be addressed by comparing the values observed for key CPs in cities of similar conditions.

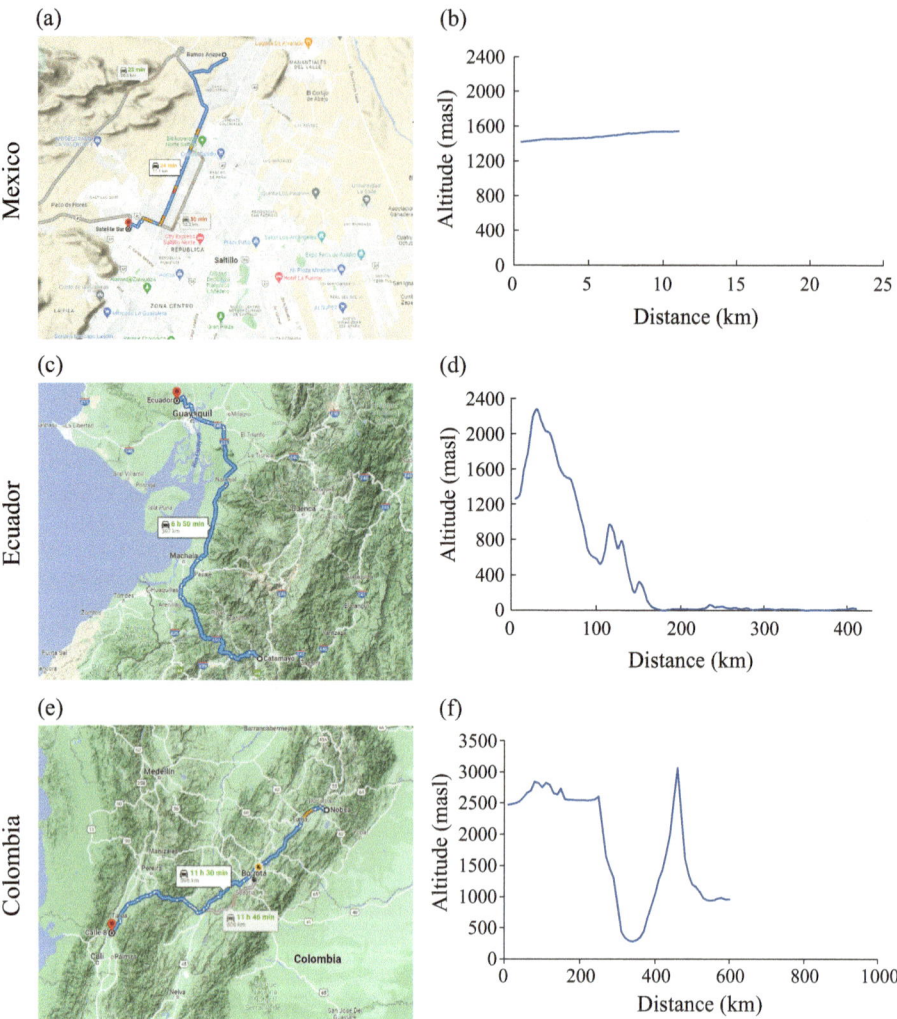

Fig. 7.1 Description of the routes where the operation of the technology described in Table 7.1 was monitored with the purpose of comparing driving patterns. Top view of the routes: **a** Saltillo-Ramos Arizpe, Mexico. **c** Catamayo-Guayaquil, Ecuador, and **e** Buga-Puerto Concordia, Colombia. Altitude variation of the route in **b** Mexico, **d** Ecuador and **f** Colombia

Table 7.3 presents some CPs of several driving patterns studied in the world. The list of CPs and reported driving patterns is incomplete. Furthermore, the list of driving patterns shown in 7.2 is dissimilar. Therefore, the driving patterns shown are not comparable to a certain extent. Some of them are for cities, and others for national wide regions. However, it can be used to highlight some relevant aspects.

Table 7.2 Description of the routes where the operation of a single technology was monitored to compare driving patterns

Parameter	Mexico	Ecuador	Colombia	Units
Location	Saltillo-Ramos Arizpe	Catamayo–Guayaquil	Buga–Puerto Concordia	–
Description	Urban	Interstate	Interstate	–
Length	21	408	506	km
Altitude range	1,420–1,560	0–2,500	200–3,500	masl
Average load transported	21	17	40	t
Trips	99	30	24	–
Total monitoring time	57.6	288	581	h
Datapoints	207,604	1,037,232	2,094,837	–

Mean speed and the percentage of idling time are important CPs because they are indicators of traffic congestion. Low speeds and high percentages of idling time are correlated with traffic congestion. These two CPs could be correlated to a high population density and a high number of cars per habitant. The number of cars per habitant can be related to citizens' high economic acquisition power. However, low speeds and high idle times can also be correlated to a poor public transportation system and or the lack of the use of information technology to improve mobility in the city. Even though some cities have reduced the maximum speed limits allowed within the city, in practice, the observable average speeds in cities are more limited by traffic than by speed limits.

Table 7.3 shows that the values for mean speed range from 16.05 to 34.30 km/h in urban cities. Notably, Mexico City and Singapore stand out for exhibiting the lowest mean speeds, indicating challenging traffic conditions. Similarly, values for idling range from 15.11 to 36%, with Europe and Nanjing being the regions that exhibit the worst conditions. Perhaps these observations are not new, and some people can argue that there is no need to go through the whole subject of driving patterns to reach a conclusion that can be just observed by visiting those cities. These figures dare tangible evidence to support authorities in their initiatives to improve mobility in their cities.

We recommend using the following metrics based on CPs to rank urban mobility (M_s) (Eq. 7.1). It ranges from 0 to 100%, with 100% being the best. It is an average value of possible mean speed (S) and percentage of idling time (I). The min and max index stand for the maximum and minimum values observed in similar regions to the one of interest or the ones described in the previous paragraph.

$$M_s = \frac{1}{2}\left(\frac{S - S_{\min}}{S_{\max} - S_{\min}} + \frac{I_{\max} - I}{I_{\max} - I_{\min}}\right) \tag{7.1}$$

Table 7.3 Comparison of different driving patterns by means of CPs

	Region	Speed (km/h)	Accel. (%)	Decel. (%)	Idling (%)	Cruise (%)	SFC (L/100 km)	KPE (m/s^2)	Density (hab/km^2)	PIB per capita ($/hab)
LATAM	Saltillo	24.07	41.65	33.72	21.12	3.50	39.51	0.34	153.6	28,400
	Ecuador	22.35	43.25	34.76	20.26	2.50	41.47	0.38	52.0	6,391
	Chile	31.04	37.05	37.78	18.88	6.30	29.00	0.18	22.0	15,355
	Bogota	19.07	40.00	36.23	23.02	0.73	47.25	0.38	4,310.0	6,642
	Mexico City	32.70	42.88	41.44	15.11	1.40	36.10	0.32	6,000.0	23,018
Europe	Istanbul	34.30	–	–	16.90	–	–	–	2,523.0	25,000
	Edinburg	18.70	34.00	34.00	28.00	4.00	–	–	1,844.0	33,033
	European	19.00	19.00	14.00	36.00	30.00	–	–		
Asia	Celje	25.50	26.00	25.00	25.00	25.00	–	–	103.9	29,750
	Hong Kong	25.00	34.50	34.20	17.80	12.00	–	–	6,668.0	48,614
	Pristina	18.42	32.47	32.98	22.42	12.13	–	–	962.0	5,340
	Chennai	22.77	30.00	27.00	19.00	24.00	–	–	26.9	
	Singapore	32.80	28.50	25.30	25.70	20.60	–	–	7,743.0	60,864
	Mashhad	20.41	37.34	37.69	21.75	3.22	–	–	9,149.0	4,669
	Nanjing	30.73	27.00	23.00	30.00	20.00	–	–	178.8	11,935
	Beijing	19.27	33.56	31.15	26.96	–	–	–	1,334.0	12,022
Africa	SriLanka	27.20	36.10	30.65	20.50	12.75	–	–	338.0	3,189
	Putian/ Fuzhou	16.05	31.13	26.72	27.15	15.00	–	–	777.0	20,270

The following family of relevant CPs is related to acceleration and fuel consumption. Average positive acceleration and deceleration are key CPs directly associated with driving aggressiveness and energy fuel consumption. Positive kinetic energy is also a relevant CP related to traffic and energy fuel consumption. In this chapter, we prefer to deal with the family of CPs related to acceleration in terms of SAPDs and the family of CPs related to fuel consumption in terms of VSPs.

Comparing driving patterns by means of SAPDs
To illustrate the conclusions that can be gained by analyzing driving patterns through SAPDs, we will use the SAPDs obtained for a single vehicle technology operated in different regions performing similar tasks (Fig. 7.2). In this case, the vehicle technology

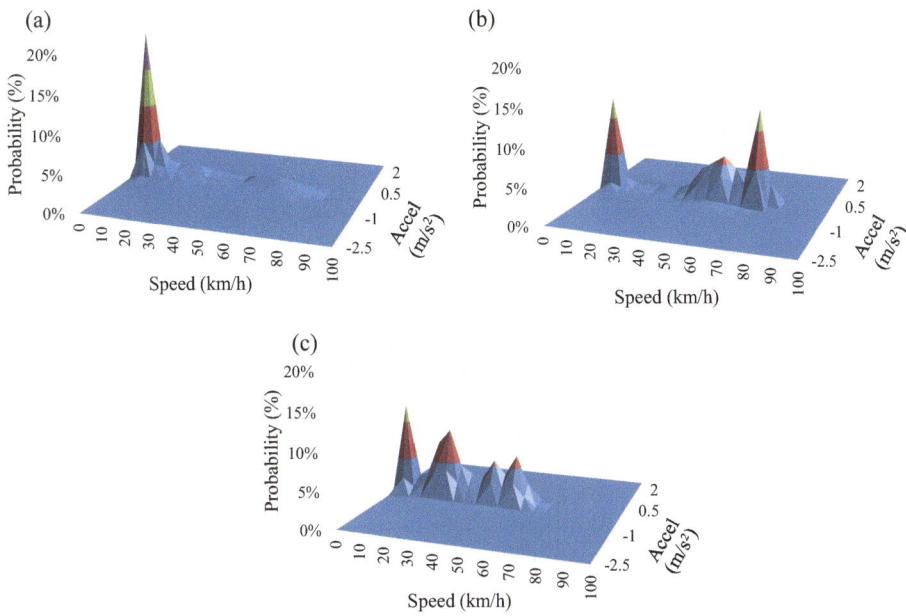

Fig. 7.2 Speed acceleration probability distribution of the same vehicle technology operating in long-distance hauling in: **a** Mexico, **b** Ecuador, and **c** Colombia. The technology monitored corresponds to tractors international prostar+ of 27.5 t powered by diesel 450 HP engines

observed is a tractor of 27 t used for long-distance hauling in Mexico, Colombia, and Ecuador.

Figure 7.2a shows the SAPD obtained after observing 99 trips of vehicles traveling between the cities of Saltillo and Ramos Arizpe (11 km) using a road network with negligible road grades covering urban areas. It shows that most of the time, the vehicles are idling or working at low speeds (<20 km/h). The drivers accelerate moderately in the range between −1.5 and 1.5 m/s². This pattern is characteristic of vehicles moving across cities.

In contrast, Fig. 7.2b shows the SAPD obtained after observing 30 trips of the same vehicle technology serving the Catamayo–Guayaquil route, which is a 408 km route ascending and descending Los Andes mountains in Ecuador from sea level up to 2500 masl. This figure exhibits three prominent peaks. The one at low speeds is characteristic of vehicles moving inside urban centers. The second peak corresponds to the vehicle ascending high road grades (7%) at medium speed (~60 km/h) and the last at moderately high speeds on flat roads. The local regulation fixed the maximum speed limit to 80 km/h.

Similar to the case of Ecuador, Fig. 7.2c shows the SAPD observed after 24 trips of the same vehicle technology operating in Colombia, serving the route from Buga to

Puerto Concordia, which is a 506 km long route crossing Los Andes Mountains, ascending from 200 masl up to 3500 masl using a road with some heavy road grades (up to 7%). In this case, the SAPD exhibits similar peaks at low (<20 km/h), medium–low (40 km/h), medium (60 km/h) and medium–high (80 km/h) speeds. Drivers again exhibit accelerations in the range of $-15 < a < 1.5$ m/s^2.

The previous description illustrates that the SAPDs capture different driving patterns. In this case, external factors highly influence the driving pattern (in this instance, topography and road conditions).

As explained in Chap. 5, SAPDs are used for constructing driving cycles by means of the Markov chain-Montecarlo method. There has been an interest in using SAPDs to evaluate drivers' performance. At first glance, this alternative does not make sense because SAPDs of different drivers are not comparable unless they operate on the same road under similar conditions and use similar vehicle technology. i.e., when the only varying factor is the human factor.

In this sense, it can be argued that the speed pattern (the frequency distribution of speeds) is mostly associated with external factors. In contrast, acceleration patterns (the frequency distribution of accelerations) are mostly associated with human factors. Therefore, the driver's performance can be evaluated by observing the acceleration frequency distribution at each speed bin in the SAPD. Aggressive drivers are those who exhibit wider acceleration frequency distributions at every speed bin. That is, assuming the vehicle technology is the same, aggressive drivers are those that exhibit high frequencies of large accelerations compared to average drivers.

We propose the following methodology to quantify this approach to evaluate drivers' aggressiveness (A, Eq. 7.2). As a starting point, let us assume we have the long-term SAPD of all drivers serving similar routes or regions using the same vehicle technology. In addition, we have the SAPD* for the driver under evaluation. Then, for a given bin speed (i), we define a threshold value for positive acceleration ($a_{(i,max)}$). As a first attempt, let us use the acceleration corresponding to the 90th percentile at the speed bin in the SAPD of all drivers. Next, the driver's aggressiveness is the summation of all acceleration frequencies ($P_{ij}{}^*$) found for the driver under evaluation above such threshold value at that bin of speeds. Next, a similar procedure is followed for the case of abrupt braking (deceleration, $a_{(i,min)}$). Finally, the procedure is repeated for each bin of speeds (i). The total summation of frequencies above critical accelerations or below critical breakings at all bins of speeds is the metric that evaluates the driver's aggressiveness. Values close to zero grade gentle drivers while values close to one identify aggressive drivers. All drivers should be listed in order according to this metric (A). Under an eco-driving program, the top 15% of aggressive drivers should undergo a training program, while the top 15% of gentle drivers should be awarded (Fig. 7.3). The definition of threshold values needs to be evaluated further through practical experience using this metric.

$$A = \sum P_{i,j}^* \text{ if } a_{i,j} > a_{i,\max} \text{ or } a_{i,j} < a_{i,\min} \qquad (7.2)$$

Fig. 7.3 Illustration of the methodology to evaluate driver's aggressiveness in eco-driving programs by comparing the SAPD* of the driver under evaluation against the **a** SAPD of eco-drivers or **b** the regions of excessive fuel consumption

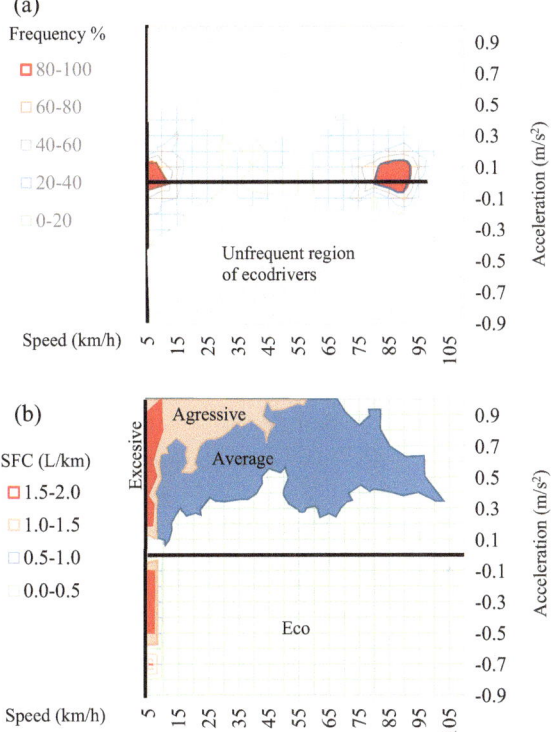

This methodology can be improved further, extending it to the case of vehicles running under different regions' roads and technology. Initially, the methodology can be applied across drivers operating in different regions and using different vehicle technologies, provided that there is an SAPD per region and technology to compare with. The need for an SAPD per region and vehicle technology is a drawback of this methodology.

The methodology can be improved further by including fuel consumption. In this case, the positive acceleration threshold value for each speed bin should be defined as the one above which specific fuel consumption is excessive. This definition implies a unique fuel consumption rate for each speed-acceleration bin, similar to the one for each torque-RPM bin, when the engine is tested under steady-state conditions.

We hypothesize that at each speed-acceleration bin, the engine works on an instant pseudo-steady state condition, where it exhibits a unique value of specific fuel consumption (L/km) and that the specific fuel consumption grows with acceleration. This hypothesis implies that for each speed-acceleration bin, there is a unique corresponding torque-RPM bin, which is not necessarily true. The presence of vehicle transmission makes the vehicle can move at a given speed with one or another transmission ratio, each of them with different levels of engine torque demand. However, vehicles may still be driven under good driving practices where the one-to-one relation between the speed-acceleration and torque-rpm bins is observed. Work is underway to evaluate this alternative of evaluating drivers' aggressiveness.

Comparison of Driving Patterns Through VSP Frequency Distributions Diagrams

Figure 7.4 shows the VSP frequency distribution observed in the operation of the vehicle technology described in Sect. 7.1 operating in long-distance hauling in Mexico, Ecuador, and Colombia. The instant VSP was calculated using Eq. 3.11.

Figure 7.4 illustrates a consistent trend across the VSP distributions obtained for the three regions examined. Correlation analyses reveal high coefficients of determination among them. The comparison base was the VSP frequency distribution observed in Mexico, yielding an $R^2 = 0.92$ and an $m = 0.55$ for the Ecuadorian VSP distributions. Similarly, comparing the Colombian VSP distribution to the Mexican VSP counterpart resulted in an $R^2 = 0.89$ and an $m = 0.64$.

The high levels of similarity suggested that the VSP distribution exhibits a universal trend, independent of variables such as vehicle size, technology, and driving style. Preliminary results contradicted this hypothesis. However, this observation triggered the dimensionless analysis detailed in the following section.

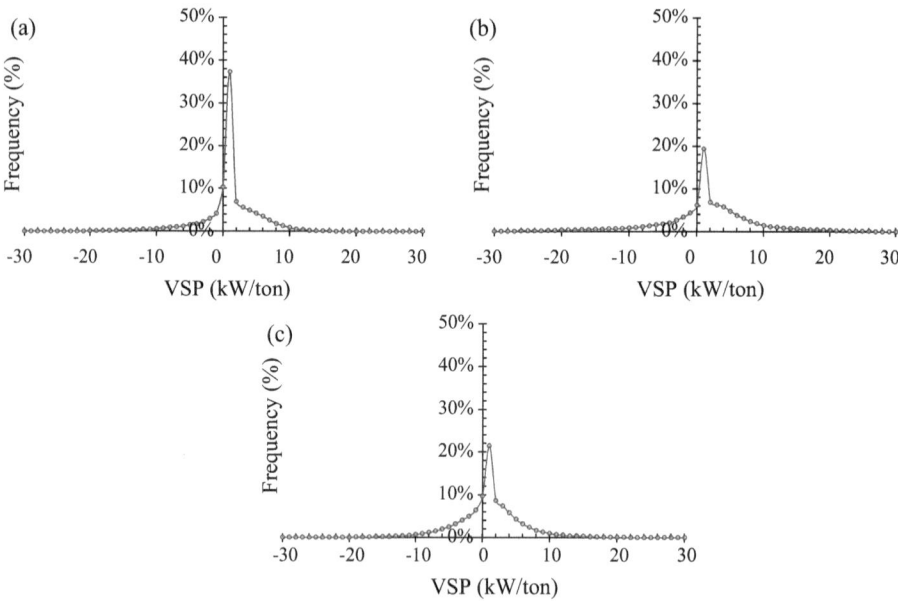

Fig. 7.4 VSP Frequency distribution observed in the operation of a single vehicle technology operating in long-distance hauling in: **a** Mexico, **b** Ecuador, and **c** Colombia. The technology monitored corresponds to tractors international prostar+ of 27.5 t powered by diesel 450 HP engines

We must highlight that the VSP is closely linked to the vehicle's fuel consumption and therefore, to tailpipe emissions. In fact, MOVES uses VSP to estimate the vehicle emissions from vehicle fleets. MOVES is the motor vehicle emissions system developed by the USEPA. Even though VSP is linearly correlated to fuel consumption, it has a poor correlation with tailpipe emissions because a single VSP can be obtained with the engine working in several working modes (combinations of torque and RPM).

Previous reasoning can be used to extract practical implications of the VSP frequency distribution diagram (f) regarding inappropriate driving techniques (*IDT*, Eq. 7.3). First, negative *VPS* are associated with energy losses through breaking or potential energy losses when the vehicle goes down the hill. Thus, in flat terrains, the integral under the *VPS* frequency distribution diagram (f) in the range $-\infty < VSP < 0$ could be used as a metric of inappropriate driving techniques. Similarly, the frequency at $VPS = 0$ is associated with idling time. Therefore, high frequencies at $VPS = 0$ are associated with inappropriate driving techniques. Finally, high frequencies at high *VPS* are related to excessive energy consumption. Therefore, the integral of the *VPS* frequency distribution (f) in the range of $VSP_{\max} < VSP < \infty$ can also be used as a metric of inappropriate driving techniques. Work should be conducted to identify recommended values for $VPS_{\max}$ and values of *IDT* typical of eco-drivers.

$$IDT = \int_{-\infty}^{0} f dVSP + f_{(VSP=0)} + \int_{VSP_{\max}}^{\infty} f dVSP \qquad (7.3)$$

7.4 Dimensional Analysis to Compare Driving Patterns

We propose an additional alternative to compare driving patterns in different regions. We suggest using dimensional analysis to identify metrics that could be used towards increasing energy efficiency in vehicles and reducing their tailpipe emissions.

Rearranging Eq. 7.3, we obtain Eq. 7.4. The left term on this equation represents the power provided by the fuel to the engine. The terms on the right represent the power associated with overcoming the forces that oppose the vehicle's motion. The first term quantifies the power associated with the drag force, the second is the rolling resistance, and the third is the gravitational force. The last term quantifies the power associated with the inertia force. That is, the power required to increase the vehicle's speed and power train (all rotational components). It represents the power consumption while the vehicle is idling. O, in Eq. 7.4, represents additional sources of power consumption, such as the power associated with the operation of air conditioning, lighting, interior amenities, etc., which will be neglected in this chapter.

$$\dot{v}_f \eta_{TD} \eta_{th} LHV \rho_f = \left[\frac{1}{2} Cd \rho_a A V^2 + f_r mg \cos(\theta) + mg \sin(\theta) + mm_f a \right] V + I + O \quad (7.4)$$

In Eq. 7.4, $\dot{v}_f$ is the fuel consumption rate expressed in L/s, η_{TD} is the mechanical efficiency of the transmission and mechanical elements from the engine to the wheel, ηth is the thermal efficiency of the diesel engine, LHV is the lower heating value of the diesel expressed in kJ/kg, and ρ_f is the density of the fuel expressed in kg/m^3. Terms in Eq. 7.4 can be reorganized as in Eq. 7.5, in the way that each term becomes dimensionless. Equations 7.5 and 7.6 define the dimensionless number that quantifies the relative contribution of each road force to vehicle fuel consumption.

$$1 = \frac{[f_r mg \cos(\theta)]v}{\eta_{TD}\eta_{th}LHV\,\rho_f\,\dot{V}_f} + \frac{\frac{1}{2}Cd\,\rho_a Av^3}{\eta_{TD}\eta_{th}LHV\,\rho_f\,\dot{V}_f} + \frac{mm_f av}{\eta_{TD}\eta_{th}LHV\,\rho_f\,\dot{V}_f}$$
$$+ \frac{[mg \sin(\theta)]v}{\eta_{TD}\eta_{th}LHV\,\rho_f\,\dot{V}_f} + \frac{I}{\eta_{TD}\eta_{th}LHV\,\rho_f\,\dot{V}_f} \tag{7.5}$$

$$1 = \Pi_r + \Pi_d + \Pi_i + \Pi_g + \Pi_o \tag{7.6}$$

As an illustration of the application of this methodology, focused on vehicle energy consumption to describe driving patterns, we calculated each Π number in Eq. 7.6 for the cases of the vehicles monitored in Mexico, Ecuador, and Colombia, described in previous sections. The Π number associated with idling was calculated as the average value of the fuel consumption rate observed. At the same time, the vehicle remained at a speed smaller than 1 km/h during the observation period. A value of 0.55, 0.92, and 1.77 mL/s were observed for the case of Mexico, Ecuador, and Colombia, respectively.

Table 7.4 shows the average values of each Π number in the three countries. It shows that almost half (48.76%) of the fuel consumed by the monitored vehicles operating in Mexico is associated with inertial forces, while in Ecuador and Colombia, the fuel consumption of the vehicles is mostly (50.66% and 61.57%) associated with supplying the potential energy needed to climb The Andes Mountains. The second energy consumer in these two last countries is again associated with overcoming the inertial forces. As expected, the fuel consumption associated with the drag and rolling resistance forces is minor in all three countries. These results demonstrate that dimensionless numbers provide relevant information on where to focus when looking for strategies to reduce fuel consumption.

For the illustrative cases described in this chapter, the dimensionless numbers show that the priority should be the reduction of the fuel consumption associated with inertial forces. These forces depend directly on the vehicle's acceleration, which is directly related to the driver's driving habits. Therefore, efforts should be concentrated on improving eco-driving techniques.

The second observation highlights a significant factor: vehicles operating on road networks with large altitude variations allocate considerable energy consumption to overcome elevation changes. Not much can be done to reduce this energy consumption except to invest in new road infrastructure with tunnels or layouts that minimize the variations in

Table 7.4 Average values of the dimensionless numbers that describe driving patterns observed in the operation of a single vehicle technology operating in long-distance hauling in Mexico, Ecuador, and Colombia. The technology monitored corresponds to tractors international Prostart+ of 27.5 t powered by 450 HP diesel engines

Name	Symbol	Definition	Region		
			Mex (%)	Ecu (%)	Col (%)
Rolling resistance	Π_r	$\dfrac{[f_r mg\cos(\theta)]\,V}{\eta_{TD}\eta_{th}LHV\rho_f\dot{v}_f}$	12.76	13.09	14.14
Drag	Π_d	$\dfrac{\frac{1}{2}Cd\,\rho_a AV^3}{\eta_{TD}\eta_{th}LHV\rho_f\dot{v}_f}$	6.80	11.77	2.83
Inertia	Π_i	$\dfrac{mm_f a\,V}{\eta_{TD}\eta_{th}LHV\rho_f\dot{v}_f}$	48.76	20.22	13.46
Gravity	Π_g	$\dfrac{[mg\sin(\theta)]\,V}{\eta_{TD}\eta_{th}LHV\rho_f\dot{v}_f}$	19.43	50.66	61.57
Idling	Π_o	$\dfrac{I}{\eta_{TD}\eta_{th}LHV\rho_f\dot{v}_f}$	12.22	4.20	7.98

the altitude of the road. Roads with constant road grades are preferable. It can also be explored alternative roads that could be even longer but with less variation in the total altitude.

This potential energy gained by the vehicles when they gain altitude can be recovered when the vehicle goes down the mountain. However, in most cases it is lost. Usually, the vehicle potential energy is converted into kinetic energy and heat in the vehicle brake system that needs to be activated to control the vehicle speed under safe values. Therefore, including a regenerative brake system that converts rotational kinetic energy into electricity could be advantageous. Nevertheless, since these vehicles are diesel-fueled, the recovered energy should be used to power the peripheral systems of the vehicle, such as air conditioning, lighting, and amenities (radio, TV, communication, etc.) within the vehicle.

7.5 Concluding Remarks

This chapter was devoted to illustrating the observations that can be obtained by comparing the driving patterns of individuals or groups of people. The discussion was focused on observations that could be used to trigger strategies oriented toward the reduction of energy consumption in the operation of vehicles. It highlighted the relevance of attempting to isolate the influence of external, technological, and human factors using CPs, SAPD, and VSP.

We described that CPs, and especially the average speed (and the speed frequency distribution) and percentage of idling time (speed = 0), are closely connected to traffic (external factor). Based on this observation, we proposed a metric to evaluate mobility.

SAPDs are mostly connected to driving styles (human factor), and therefore, they have been used to evaluate drivers' aggressiveness. We proposed a methodology to quantify driving aggressiveness and how to use it in eco-driving programs.

Finally, the VSP frequency distribution is closely connected to fuel consumption. We added to the concept of VSP the use of dimensional analysis and developed dimensionless numbers for the fuel consumption associated with rolling resistance, drag, gravity, inertia, and idling.

We illustrated these concepts through an example. Results were obtained from deploying previous analyses to the data obtained by monitoring via telemetry a single vehicle technology (tractor truck of 27 t) used for long hauling in different regions (Mexico, Ecuador, and Colombia). Comments were added based on the results obtained to design strategies to reduce fuel consumption in each region.

Case 1: Chile

8

Franco Quezada, Felipe Vásquez, Oscar S. Serrano-Guevara, and José I. Huertas

Abstract

Studies report differences in performance for vehicles of similar technology, suggesting that variables affect energy consumption, such as operating conditions. This work seeks to determine the energy consumption of a fleet of 8 diesel buses under real operating conditions in Chile. This fleet was monitored using a commercial telemetry system for over three months, with data obtained at 1/15 Hz of location, speed, engine RPM, and energy consumption. As a result, an average consumption of 0.430 L/km and a weighted average consumption of 0.417 L/km were obtained for diesel buses. In this case, accelerations and variations in altitude were the variables that most affected energy consumption. Finally, a mixed typical driving cycle (urban and highway) was determined.

F. Quezada · F. Vásquez
Mechanical Engineering Department, University of Concepcion, Concepción, Chile
e-mail: fnquezadab@gmail.com

F. Vásquez
e-mail: fevasquez@udec.cl

O. S. Serrano-Guevara · J. I. Huertas (✉)
Sustainable Energy Research Group, Tecnologico de Monterrey, Monterrey 64849, México
e-mail: jhuertas@tec.mx

O. S. Serrano-Guevara
e-mail: oserrano@tec.mx

© The Author(s), under exclusive license to Springer Nature Switzerland AG 2025   117
J. I. Huertas (ed.), *Fundamentals of Driving Patterns and Driving Cycles*, Synthesis
Lectures on Mechanical Engineering, https://doi.org/10.1007/978-3-031-76863-7_8

8.1   Introduction

The agreement to achieve net zero emissions by 2050 has motivated the global elimination or reduction of fossil fuel consumption [1]. In transportation, the choice has been made to electrify the power train system of the vehicles and increase its energy efficiency. In particular, the components, vehicle, and fleet levels need intervention to improve efficiency. The first two levels can be analyzed in a laboratory. In contrast, the fleet must necessarily be analyzed during real operation since energy consumption varies significantly with the operating conditions of the vehicles.

To generate robust energy efficiency strategies, it is necessary to understand the real sources of energy consumption during real operations. A first approach can be achieved through the characteristic parameters (CPs). For example, high average accelerations, increased top speeds, and long idle time imply greater energy consumption. A second method consists of quantifying the energy consumption of each resistant force of the vehicle (aerodynamic drag, rolling, inertial, and weight component) and of the idle periods, to prioritize the energy consumption and identify the most significant ones. With either of the two methods, it is necessary to know the speeds, accelerations, and altitude variations during the entire operating time, along with the vehicle parameters.

Vehicle fleet telematics companies are in a strategic position to meet this objective because they work with the operation data of each vehicle within the fleet. A telematics company seeks to implement this analysis to advise fleet operators on improving their operation's energy efficiency.

8.2   Data Gathering and Processing

In collaboration with Inway sPA, the normal operation of 8 buses, owned by a private people transportation company with presence throughout the country, was monitored. The primary monitoring services are the status of the vehicle batteries, evaluation and control of driving risk, and integration of peripheral elements in the cabin. The objective is to implement a more exhaustive monitoring process that allows proposing improvements aimed at energy efficiency.

The main monitored variables were engine RPM, cumulative fuel consumption, speed, and position; the used device is described in [2], and its reporting frequency is 1/15 Hz which is one data point every 15 s. To obtain the 1 Hz reported data, linear interpolation was used [3]. The specifications for the 8 buses monitored for 4 months are shown in Table 8.1, while Table 8.2 shows the details of the monitoring campaign, where we highlighted that 8.57 M second-by-second data points were collected. Finally, an example of route tracking is shown in Fig. 8.1.

Table 8.1 Technical specifications of monitored vehicles for this case

Manufacturer	Model	Model year	Qty	GVWR (t)	Engine displacement (L)	Engine power (kW @ rpm)	Engine torque (Nm @ rpm)
Mercedes Benz	0 500 RS 1936 4 × 2	2021	8	19.6	12.9	345 @ 1,800	2,200 @ 1,100

Table 8.2 Details of monitoring campaign

Detail	Value
Monitoring period (months)	4
Frequency (Hz)	1/15
Since, to	Nov 2022–Feb 2023
No. of vehicles (#)	8
Collected data (#)	8.57 M
Distance (km)	76,763.54
Used fuel (L)	32,056.45
Specific fuel consumption (L/km)	0.417

As data quality analysis process, the position points (latitude and longitude) were plotted in a geographical information software (as shown in Fig. 8.1), and some missing points were detected and corrected. The next step was to take valid cumulative consumption data; for this, an operating range was determined through the standard deviation of this variable, as reported in [4].

After data quality analysis, the mentioned linear interpolation process was applied, and then the main results, such as characteristic parameters, average energy consumption, and typical driving cycle, were determined.

8.3   Results

Table 8.3 summarizes the main characteristic parameters for each monitored vehicle. Urban driving conditions are mainly distinguished by the low average speeds, around 25–30 km/h, and high idle times, over 40%. On the other hand, it is also possible to distinguish cautious and aggressive driving styles. For the gentle case, average accelerations of $[-0.4, 0.4]$ m/s^2 were observed, like in buses 808 and 849, while for aggressive driving, average accelerations were observed in the range of $[-0.6\ 0.6]$ m/s^2, for buses 847 and 851.

Fig. 8.1 Sampled bus route through center-south from Chile

In addition, Table 8.3 shows that buses 808 and 847 show extra-urban or highway driving conditions, but the first one, which exhibit a gentle driving style (lower average accelerations), has a lower fuel consumption (0.299 L/km) than the second one, which shows higher average acceleration (0.316 L/km).

Figure 8.2a shows the specific fuel consumption (SFC) of the different vehicles, where an average consumption of 0.397 L/km and a weighted average of 0.417 L/km were obtained. This value is slightly higher than that reported in the literature for similar vehicles, like 0.324 L/km in [5], 0.37 L/km in [6], and 0.384 L/km in [7]. The typical driving cycle for bus operating conditions in Chile is shown in Fig. 8.2b. This cycle was determined using the energy-based Micro-trip method (EBMT).

Table 8.3 Main characteristic parameters observed for each monitored bus

Characteristic parameter/bus id	808	845	847	849	851	852	853	860
Max speed (km/h)	103	113	104	100	101	104	102	109
Average speed (km/h)	42.98	29.81	43.16	18.53	24.73	32.56	31.85	33.99
Sd of speed (km/h)	41.48	27.42	35.01	17.51	27.45	29.65	28.48	29.45
Average positive acceleration (m/s^2)	0.98	1.21	1.45	1.15	1.42	1.5	1.22	0.99
Average negative acceleration (m/s^2)	−1.01	−1.25	−1.42	−1.55	−1.36	−1.28	−1.38	−1.06
Average positive acceleration (m/s^2)	0.31	0.49	0.66	0.41	0.62	0.5	0.47	0.5
Average negative acceleration (m/s^2)	−0.37	−0.54	−0.74	−0.43	−0.67	−0.55	−0.53	−0.55
Sd of positive acceleration (m/s^2)	0.32	0.28	0.31	0.26	0.29	0.25	0.24	0.31
Idling percentage (%)	35.6	44.54	30.96	48	43.21	41.3	42.99	39.6
Acceleration percentage (%)	10.81	12.95	14.26	9.68	12.47	12.64	11.89	12.18
Deceleration percentage (%)	9.96	12.19	13.37	9.69	11.73	12.24	11.24	11.78
Cruising time (%)	43.64	30.32	41.4	32.63	32.59	33.82	33.88	36.43
SFC (L/km)	0.299	0.430	0.316	0.549	0.375	0.398	0.422	0.421

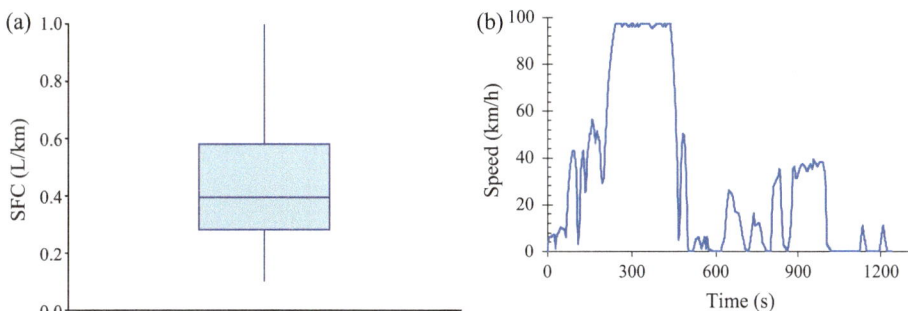

Fig. 8.2 **a** Specific fuel consumption (SFC), and **b** typical driving cycle for buses operating in Chile

Table 8.4 Characteristic parameters and absolute relative differences of the driving cycle and the driving pattern observed in Chile

Characteristic parameter	Driving cycle	Driving pattern	Relative difference
Time (s)	1,236	8,571,611	NR
Max speed (km/h)	97	113	9%
Average speed (km/h)	31.04	30.35	2%
Sd of speed (km/h)	34.81	32.24	8%
Maximum acceleration (m/s^2)	0.97	1.71	43%
Minimum acceleration (m/s^2)	−1.42	−2.08	32%
Average positive acceleration (m/s^2)	0.39	0.37	5%
Average negative acceleration (m/s^2)	−0.46	−0.14	228%
Sd of positive acceleration (m/s^2)	0.28	0.21	32%
Idling percentage (%)	18.85%	20%	7%
Acceleration percentage (%)	22.09%	14%	63%
Deceleration percentage (%)	18.53%	13%	41%
Cruising time (%)	40.53%	53%	24%
SFC (L/km)	0.379	0.418	9%

The main characteristic parameters of this cycle, as well as those of the driving pattern and their relative difference, are shown in Table 8.4, where it is observed that the main parameters show absolute relative differences of less than 10%.

8.4   Conclusions

The buses reported in this case have a specific fuel consumption slightly higher than others reported in the literature under real operating conditions, which is mainly due to the local conditions of use of these vehicles, which reaffirms that results cannot be generalized without first describing the real operating conditions, like average road grade, altitude, and main characteristic parameters.

When comparing the three main characteristic parameters, along with energy consumption, the determined driving cycle meets the minimum relative difference with the driving pattern. This shows a high percentage of cruising due to highway operation and another mixed part: accelerations, decelerations, and idle, where urban driving is observed (Table 8.4).

Author Contributions Francisco Quezada: Data curation, Methodology, Investigation, Formal analysis, Writing—original draft. **Felipe Vásquez**: Supervision, Investigation, Formal analysis, Writing—review and editing, Data acquisition. **Oscar Serrano-Guevara**: Conceptualization, Methodology, Data curation, Investigation, Formal analysis, Writing—review and editing. **José I. Huertas**: Conceptualization, Methodology, Supervision, Investigation, Formal analysis, Writing—review and editing, funding acquisition.

References

1. UN for a Livable Climate: Net-Zero Commitments Must Be Backed by Credible Action. https://www.un.org/en/climatechange/net-zero-coalition. Accessed on 15 February 2024
2. Inway Tecnología. https://www.inway.cl/tecnologia/. Accessed on 15 February 2024
3. Huertas, J.I., Serrano-Guevara, O., Díaz-Ramírez, J., Prato, D., Tabares, L.: Real vehicle fuel consumption in logistic corridors. Appl. Energy **314** (2022). https://doi.org/10.1016/j.apenergy.2022.118921
4. Lin, L.: Cleaning Data the Chauvenet Way (2017)
5. Ghaffarpasand, O., Talaie, M.R., Ahmadikia, H., Khozani, A.T., Shalamzari, M.D., Majidi, S.: Real-world assessment of urban bus transport in a medium-sized city of the middle east: driving behavior, emission performance, and fuel consumption. Atmos. Pollut. Res. **12**, 113–124 (2021). https://doi.org/10.1016/j.apr.2021.02.004
6. Giraldo, M., Huertas, J.I.: Real emissions, driving patterns and fuel consumption of in-use diesel buses operating at high altitude. Transp. Res. D Transp. Environ. **77**, 21–36 (2019). https://doi.org/10.1016/j.trd.2019.10.004
7. Serrano-Guevara, O.S., Huertas, J.I., Quirama, L.F., Mogro, A.E.: Energy efficiency of heavy-duty vehicles in Mexico. Energies (Basel) **16** (2023). https://doi.org/10.3390/en16010459

Case 2: Loja, Ecuador

Jairo Castillo-Calderón and Daniel Cordero-Moreno

Abstract

A typical driving cycle represents a distinctive way of driving in a city or highway, considering vehicle technology, traffic characteristics, climatic and geographical characteristics, and driving style. These driving cycles are of great importance, among other purposes, to adequately plan the development of a city, the development of technology for new automobiles, and inventories of polluting emissions and energy consumption. Under this context, the representative driving cycle of an electric vehicle used as a taxi in Loja, Ecuador, is determined. This driving cycle was defined to specify the behavior and driving pattern of the city's electric vehicle used as public transportation and its intrinsic relationship with energy consumption. A speed-time series of 548 s, 36% acceleration time, and 0.57 kWh traction energy consumption was determined as a representative driving cycle with the deterministic methodology of Minimum Weighted Differences of Characteristic Parameters (MWD-CP).

J. Castillo-Calderón (✉)
Facultad de la Energía, las Industrias y los Recursos Naturales No Renovables, eX-MoVeT
Research Group, Universidad Nacional de Loja, 110103 Loja, Ecuador
e-mail: jdcastilloc@unl.edu.ec

D. Cordero-Moreno
Facultad de Ciencia y Tecnología, ERGON, Centro de Investigación y Desarrollo en Ingeniería
Automotriz, Universidad del Azuay, 010204 Cuenca, Ecuador
e-mail: dacorderom@uazuay.edu.ec

9.1   Introduction

Since the early 1960s, vehicles have been tested to meet emissions standards using standardized tests, known as test cycles or driving cycles. A driving cycle can be seen as a sequence of test points, each with a defined vehicle speed to be followed by the vehicle under study; these test points are divided into time steps, typically seconds, during which acceleration and deceleration are assumed constant [1]. That is, driving cycles are typically defined in terms of vehicle speed as a function of time [2].

The most important quality of a driving cycle is the degree to which it reflects actual driving behavior, that is, its representativeness of the actual driving pattern. Under this context, the definition of the Typical Driving Cycle (TDC) or local driving pattern was born, understood as a time series of speeds that represents the average driving pattern of drivers over the real traffic conditions in a given city or region [3, 4]. The characteristics of driving cycles in specific areas are diverse because they significantly differ in road structure, types of roads and vehicles, traffic conditions, driving culture, geographical characteristics, economic level, and city scale [5]. Therefore, each region must have a TDC [6].

TDCs are used for vehicle powertrain design to evaluate vehicle fuel consumption and emissions compliance before entering a country's automotive market [7]. In addition, they can compare vehicle performance, calculate energy consumption, develop emissions inventories, and evaluate the impact of traffic [8]. In recent years, due to the deployment of hybrid and pure electric vehicles, new TDCs have been developed to evaluate energy management, batteries, energy storage capacity, and range [9].

According to the National Institute of Statistics and Censuses (INEC), in 2020, 637 electric vehicles (EVs) were registered in Ecuador, equivalent to 0.027% of the national vehicle fleet of more than 2.3 million motor vehicles. Precisely, in 2017, the southern city of Loja pioneered the introduction of 51 EVs to public transport in the form of taxis, forming the company ELECTRI LOJA ECOLOSUR. To this end, the government and the local municipality proposed financial incentives for current vehicle owners to purchase the units. The truth is that, in Ecuador, from the energy and environmental point of view, the transition from vehicles operating with fossil fuels to EVs results in a great challenge and a significant need, considering that transportation is the sector with the highest energy consumption, showing sustained growth over time [10].

In the decision to opt for vehicles with an electric propulsion system to achieve a sustainable future, much of the research has focused on the electrification of passenger cars since they are the ones that contribute the most to transport emissions with 3 GtCO$_2$ in 2020, equivalent to 41.10% [11]. Therefore, this class of vehicles presents an opportunity to achieve enormous benefits in the fight against the growing effect of GHGs [12]. Thus, this case study's objective is to define the TDC of an electric taxi, considering the unit in service.

9.2   Case Description

The Minimum Weighted Differences of Characteristic Parameters (MWD-CP), a methodology developed by [3], is used to obtain the TDC. It is emphasized that the definition of the TDC focuses on the rides made by the electric taxi during the month. It excludes from this analysis when the taxi circulates without passengers. The objective is to choose among all the sampled rides the one that represents them, expressing each ride in terms of characteristic parameters or performance values.

P_{ij} is defined as the value of parameter i obtained for cycle j. First, the arithmetic mean of each parameter, $\overline{P}_i$, is calculated for all the sampled cycles. The second step consists of comparing each characteristic parameter to the mean value of the same parameter for all the sampled cycles, $|P_{ij} - \overline{P}_i|$, and summing the differences obtained for each parameter. However, some parameters are more relevant than others. Therefore, the sum of the differences should be weighted according to the relevance of each parameter in determining, for example, the energy demand, as in [13] where the proposed characteristic parameters are the energies demanded by the types of loads and their weight will be the percentage of their contribution to the total energy demand. Finally, as described in Eq. 9.1, the cycle, or trip, with the smallest sum of weighted differences, is selected to represent all cycles in the sample and, therefore, as the TDC.

$$C = Arg\left\{\min_j\left(\sum w_i|P_{ij} - \overline{P}_i|\right)\right\} \tag{9.1}$$

The experimental scenario comprises the free travel of the electric taxi in the urban area of Loja, Ecuador, located at an average altitude of 2060 m above sea level. This information was considered in the calculation of air density. The electric taxi used in the research is the KIA SOUL EV, equipped with an AC permanent magnet synchronous motor with a maximum torque and power of 285 Nm and 81.4 kW, respectively. It also has a lithium-ion polymer battery bank with a storage capacity of 27 kWh. To obtain the characteristic parameters, (i.e., the energies of the various forces opposing the vehicle's motion), the longitudinal vehicle dynamics are used, with the variables summarized in Table 9.1.

Table 9.1 Variables of the EV's longitudinal dynamics

Variable	Description	Value	Unit
f_r	Rolling resistance coefficient	0.017	–
C_d	Aerodynamic drag coefficient	0.35	–
A	Frontal area vehicle	2.3	m^2
ρ_a	Air density	0.88	k g/m^3

9.3   Data Gathering and Processing

The unit is monitored for one month, in daily workdays, acquiring in real time the speed and geographical variables through the OBDII port of the KIA SOUL EV, using a data logger device, model OBDLink MX+, which includes GPS, at a sampling rate of 1 Hz. The reading and storage of variables are performed with a program code developed in Labview. In addition, the initial and final time of each run is recorded, and the number of passengers is counted. This allows for obtaining a variable mass of the unit in service. That is, when the taxi circulates without passengers, the value of the mass of the taxi is considered to be the empty weight of the vehicle, which is 1492 kg, plus the weight of the driver, which is 70 kg. Whereas, if the taxi circulates with a ride, the mass of the unit will vary according to the number of passengers. NTE INEN 1323:2009 stipulates that the mass of an occupant is 70 kg. EU regulation 646 is used to establish the road gradient to be used in the vehicle dynamic model. It considers four sequential processes of altitude smoothing and two of road gradient smoothing. The smoothing begins by taking the vehicle's geographical variables—latitude, longitude, and instantaneous altitude—as a starting point. Figure 9.1 shows the final road gradient smoothing, with a maximum value of 0.28 radians, when compared with the road gradient obtained with the GPS Visualizer online application, where the monitored values of latitude and longitude of the vehicle are entered.

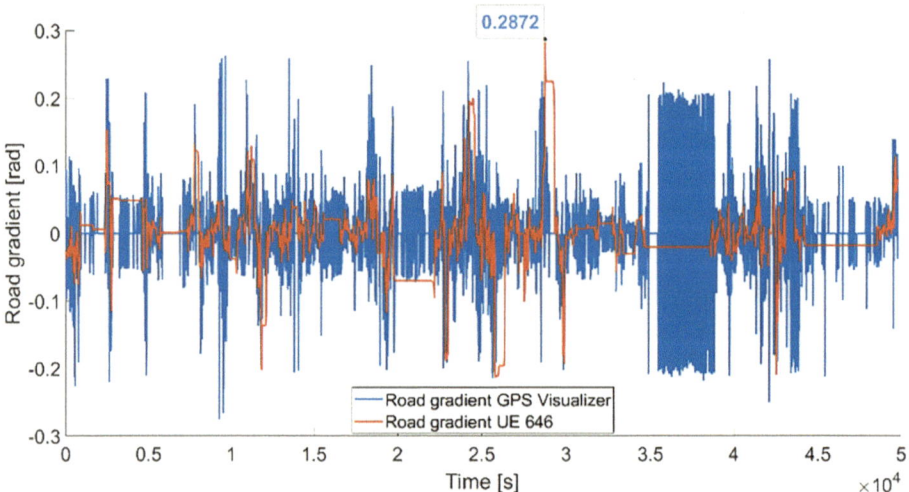

Fig. 9.1 Road gradient of the EV's route on day 3

9.4   Results

The histogram in Fig. 9.2 represents the frequency distributions of the rides classified by time of day. The electric taxi performed 660 monthly rides between 03:00 am and 7:00 pm. The results show that the mean of the rides was set at 09:25 am. The data distribution exhibits a considerable predominance of rides in the morning hours, suggesting higher labor productivity. The median value indicates that half of the rides occur before 9:30 am. Per month, the taxi performed more rides, 75 in total, in the 8:00–9:00 am time slot, and only one ride in the 6:00–7:00 pm time slot.

Table 9.2 summarizes the results derived from the deterministic MWD-CP methodology. The TDC corresponds to ride 5 of day 11, with weighted differences of 0.25, the lowest among all the 660 rides. Of the characteristic parameters, the percentage of energy demanded by inertia stands out in first place, with 49.48%, and in last place, the percentage of energy requested by aerodynamic drag, with 3.48%.

Table 9.3 shows the parameters related to the TDC, corresponding to ride 5 of day 11. Among them, it is highlighted a ride duration of 9 min and 8 s, where 3.4 km were circulated, with one passenger on board. The high proportion of the acceleration and

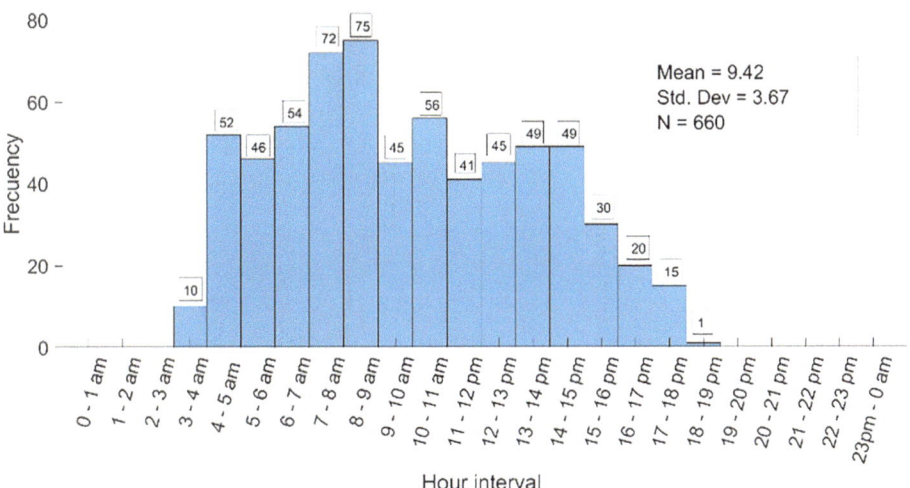

Fig. 9.2 Histogram of monthly rides per hour

Table 9.2 Traction energy consumption results from rides

Day	Ride	%EF$_d$(+)	%EF$_f$(+)	%EF$_c$(+)	%EF$_a$(+)	Σ
11	5	3.48	27.82	19.22	49.48	0.25

Table 9.3 Summary of TDC parameters

Parameter	Value	Unit
Total time	548	[s]
Total distance	3.4	[km]
Average speed	22.32	[km/h]
Maximum speed	45	[km/h]
Proportion of time standing	22.40	[%]
Proportion of cruising	16.58	[%]
The proportion of time accelerating	35.88	[%]
The proportion of time decelerating	25.14	[%]
Energy consumption	0.57	[kWh]
Specific energy consumption (SEC)	16.76	[kWh/100 km]

deceleration state of the built TDC can be associated with traffic congestion conditions in the urban area where the electric taxi circulates.

The typical driving cycle constructed for the electric taxi by the proposed method is shown in Fig. 9.3a. At the same time, Fig. 9.3b represents the speed-acceleration frequency distribution (SAFD) of the established real-world driving cycle.

From Fig. 9.3b, it can be seen that the velocity probability is highest in the range of 0 to 10 km/h, and the acceleration is mainly distributed between −0.5 to 0.5 m/s². These results indicate that the electric taxi starts and stops several times, accelerations and decelerations are frequent, and the velocity is low, which agrees with the findings of [14]. This condition is typical of a vehicle driving in an urban area with considerable traffic and many traffic lights.

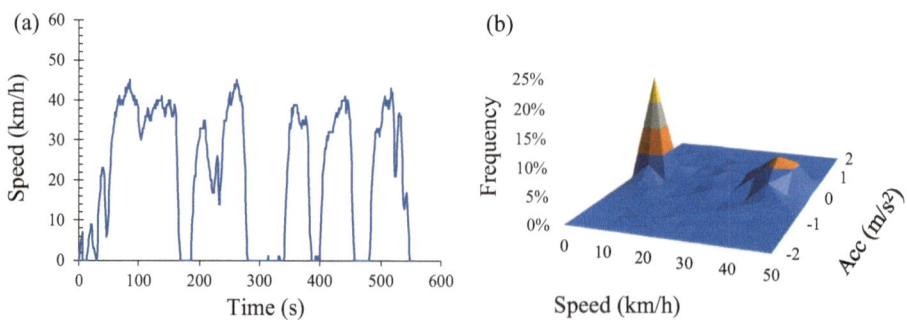

Fig. 9.3 EV's driving pattern for Loja city, **a** typical driving cycle, and **b** speed acceleration frequency distribution (SAFD) of the driving cycle

9.5 Conclusions

This study aimed to define a typical driving cycle of an electric taxi in an Andean city, considering the unit in service, that is., while circulating with passengers, using the deterministic methodology MWD-CP. 660 rides of an electric taxi were monitored in a month. The TDC corresponds to ride 5 on day 11, with weighted differences of 0.25 and a traction energy consumption of 0.57 kWh. The characteristic parameters used were the energies of the various forces that oppose the vehicle's movement, where the percentage of energy demanded by inertia had the most significant prominence, with 49.48%. The TDC lasted 9 min and 8 s, where the electric taxi circulated 3.4 km with a passenger on board. High proportions of the acceleration and deceleration states of the TDC were obtained, with 33.88% and 25.14%, respectively, associated with real-world driving conditions in the city's urban area, where traffic and traffic lights have a predominant role.

Author Contributions **Jairo Castillo-Calderón**: Data curation, Conceptualization, Investigation, Formal analysis, Writing—original draft. **Daniel Cordero-Moreno**: Conceptualization, Methodology, Supervision, Formal analysis, Writing—review and editing.

References

1. Giakoumis, E.G.: Driving and Engine Cycles. Springer International Publishing, Cham (2017). ISBN: 978-3-319-49033-5
2. Barlow, T., Latham, S., McCrae, I.S., Boulter, P.: A reference book of driving cycles for use in the measurement of road vehicle emissions. TRL Published Project Report (2009)
3. Huertas, J.I., Díaz, J., Cordero, D., Cedillo, K.: A new methodology to determine typical driving cycles for the design of vehicles power trains. Int. J. Interact. Des. Manufact. (IJIDeM) **12**, 319–326 (2018). https://doi.org/10.1007/s12008-017-0379-y
4. Zhao, X., Ye, Y., Ma, J., Shi, P., Chen, H.: Construction of electric vehicle driving cycle for studying electric vehicle energy consumption and equivalent emissions. Environ. Sci. Pollut. Res. **27**, 37395–37409 (2020). https://doi.org/10.1007/s11356-020-09094-4
5. Knez, M., Muneer, T., Jereb, B., Cullinane, K.: The estimation of a driving cycle for Celje and a comparison to other European cities. Sustain. Cities Soc. **11**, 56–60 (2014). https://doi.org/10.1016/j.scs.2013.11.010
6. Wager, G., Whale, J., Braunl, T.: Driving electric vehicles at highway speeds: the effect of higher driving speeds on energy consumption and driving range for electric vehicles in Australia. Renew. Sustain. Energy Rev. **63**, 158–165 (2016). https://doi.org/10.1016/j.rser.2016.05.060
7. Giraldo, M., Quirama, L.F., Huertas, J.I., Tibaquirá, J.E.: The effect of driving cycle duration on its representativeness. World Electr. Veh. J. **12** (2021). https://doi.org/10.3390/wevj12040212
8. Achour, H., Olabi, A.G.: Driving cycle developments and their impacts on energy consumption of transportation. J. Clean. Prod. **112**, 1778–1788 (2016). https://doi.org/10.1016/j.jclepro.2015.08.007

9. Berzi, L., Delogu, M., Pierini, M.: Development of driving cycles for electric vehicles in the context of the city of Florence. Transp. Res. D Transp. Environ. **47**, 299–322 (2016). https://doi.org/10.1016/j.trd.2016.05.010

10. Sierra, J.C.: Estimating road transport fuel consumption in ecuador. Energy Policy **92**, 359–368 (2016). https://doi.org/10.1016/j.enpol.2016.02.008

11. IEA: Energy Technology Perspectives (2020)

12. Moeletsi, M.E.: Socio-economic barriers to adoption of electric vehicles in South Africa: case study of the Gauteng province. World Electr. Veh. J. **12**, 167 (2021). https://doi.org/10.3390/wevj12040167

13. Cordero-Moreno, D., Davalos, D., Coello, M., Rockwood, R.: Proposed criteria to determine typical vehicular driving cycles using minimum weighted differences. In: Urban Transport XXIII. WIT Transactions on the Built Environment, vol. 176, pp. 329–337 (2017)

14. Qin, X., Yu, K., Li, H., Dai, F., Liu, H., Yang, H., Ye, J., Zhu, H.: Development of a one-day driving cycle for electric ride-hailing vehicles. Transp. Res. D Transp. Environ. **89**, 102597 (2020). https://doi.org/10.1016/j.trd.2020.102597

Case 3: Cuenca, Ecuador

10

Daniel Cordero-Moreno, Efrén Fernández, Danilo Dávalos, and Freddy Vázquez

Abstract

Vehicular driving cycles are speed vs. time diagrams that depict typical driving patterns within a specific city or region. These diagrams are currently used to assess fuel consumption and exhaust emissions and to configure the powertrains of both new and existing vehicles. Establishing a local driving cycle, including an altitude profile, is imperative for the latter application. One of the presently employed techniques involves analyzing a sample of trips using specific parameters such as average speed, maximum acceleration, and idling time. Subsequently, minimum weighted differences are applied to derive the typical driving cycle. In this study, a methodology based on the vehicle's energy demand to overcome drag force (Fd), rolling resistance (Rx), inertia (Ri), and road gradient resistance (Rg) was used. This methodology has been applied to obtain typical Cuenca driving cycles under urban conditions.

D. Cordero-Moreno (✉) · E. Fernández · D. Dávalos · F. Vázquez
ERGON, Automotive Engineering Research and Development
Center, Universidad del Azuay, 010204 Cuenca, Ecuador
e-mail: dacorderom@uazuay.edu.ec

E. Fernández
e-mail: efernandez@uazuay.edu.ec

D. Dávalos
e-mail: danilodava_7@hotmail.com

F. Vázquez
e-mail: fnvazquez@uazuay.edu.ec

J. I. Huertas (ed.), *Fundamentals of Driving Patterns and Driving Cycles*, Synthesis
Lectures on Mechanical Engineering, https://doi.org/10.1007/978-3-031-76863-7_10

10.1 Introduction

Acquiring the altitude and speed profiles is crucial to obtaining a driving cycle for power-train configuration through simulation. The urban case study focuses on Cuenca, Ecuador, situated in the southern region at an approximate altitude of 2,500 meters above sea level. Covering an area of nearly 72 km^2, Cuenca experiences a mean temperature of around 15°C. This study collected data for the urban case using a GPS receiver installed in a taxi.

10.2 Case Description

For this case study, an extensive dataset capturing the entire city's driving dynamics was acquired using a taxi equipped with a GPS unit, providing the parameters detailed in Table 10.1. Over 30 days, the taxi was actively monitored during regular usage. Valid trips, defined as journeys with passengers from point A to point B within the city, were considered for analysis.

Trips with passengers were considered valid based on the continuously collected GPS data; taximeter information was employed to filter trips with a passenger. In the interstate case, several trips were monitored before applying the weighted differences method, using energies from road load, to obtain the driving cycle [2].

Upon identifying valid trips, the wheel power demand (P_x) was calculated using Equation 10.1, and the wheel energy demand (E_x+) was determined through Eq. 10.2. It is noteworthy that only positive energy was considered, as any negative energy represents

Table 10.1 GPS specifications [1]

Parameter	Value
Update rate	1/s, continuous
GPS accuracy	
Position	<10 m, typical
Velocity	0.05 m/s steady state
DGPS (WAAS) accuracy	
Position	<5 m, typical
Velocity	0.05 m/s steady state
Electronic compass feature	
Accuracy	±2° with proper calibration (typical); ±5° extreme northern and southern latitudes
Altimeter feature	
Resolution	1 foot
Range	−2,000 to 30,000 feet

wasted energy due to the absence of an energy recovery system in the vehicle.

$$P_x = F_x * V \tag{10.1}$$

$$E_x+ = \int P_x * dt \tag{10.2}$$

For each force, the power and energy demands were calculated, encompassing drag (Fd), rolling resistance (Rx), road grade (Rg), and inertia (Ri). Equations 10.3 and 10.4 express the power and energy demands associated with each force a vehicle must overcome.

$$P_i = F_i * V \tag{10.3}$$

$$E_i+ = \int P_i * dt \tag{10.4}$$

This comprehensive analysis enables a detailed understanding of the power and energy demands concerning the specific forces encountered during urban driving.

10.3 Data Gathering and Processing

In Cuenca, the taximeter and GPS were meticulously synchronized to determine each valid trip precisely, as illustrated in Fig. 10.1. A comprehensive dataset comprising 512 valid trips was monitored for further analysis.

For each of the 512 valid trips, energy demand was calculated. Vehicle information crucial for computing values in Eqs. 10.1–10.4 is outlined in Table 10.2.

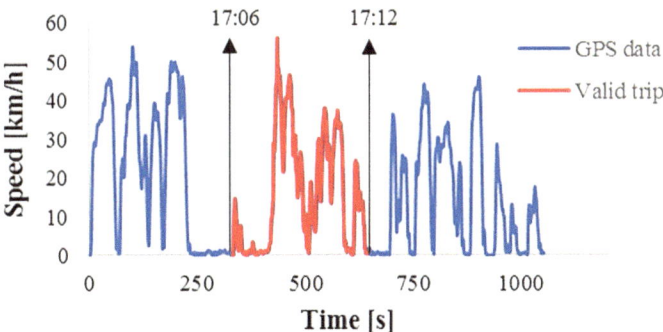

Fig. 10.1 Valid trips from GPS data

Table 10.2 Vehicle and ambient parameters

Parameter	Symbol	Value	Unit
Speed	V	From each trip	[m/s]
Acceleration	a	From each trip	[m/s^2]
Road grade	θ	From each trip	[rad]
Mass	M	1070	[kg]
Frontal area	A	1.897	[m^2]
Drag coefficient	Cd	0.33	[–]
Rolling resistance coefficient	fr	0.017	[–]
Air density	ρ	0.91	[kg/m^3]
Gravity	g	9.81	[m/s^2]

Table 10.3 Average energy demand per trip

Energy demand due to each load	Symbol	Value	Units
Energy due to inertia	E_{Fd}	326.2 ± 15.64	[Wh]
Energy due to rolling resistance	E_{Rx}	218.74 ± 10.46	[Wh]
Energy due to road grade	E_{Rg}	175.68 ± 12.52	[Wh]
Energy due to drag force	E_{Ri}	34.64 ± 3.11	[Wh]

Utilizing the data from Table 10.2 and applying Eqs. 10.1–10.4, the energy due to each type of load was computed for every trip. The results of the 512 trips, where energy demand per trip was analyzed with a 95% confidence interval, are presented in Table 10.3.

10.4   Results

With the Minimum Weighted Differences for Characterization Parameters (MWD-CP) methodology [3], it is imperative to determine characteristic parameters and their corresponding weights. For this study, the proposed characteristic parameters are the energies demanded by different types of loads, and their associated weights are defined as the percentage contribution to the total energy demand. These weights are visually represented in Fig. 10.2.

Using the weights presented in Fig. 10.2, a typical driving cycle (Fig. 10.3a), including an altitude profile (Fig. 10.3b), was selected. The driving cycle consists of a 716 s

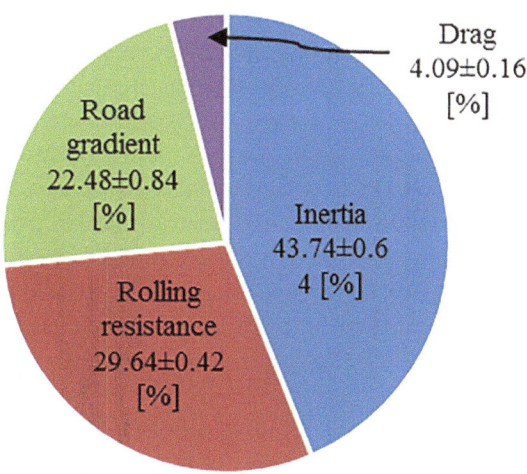

Fig. 10.2 Contribution of each vehicle load in total energy demand (per trip) [4]

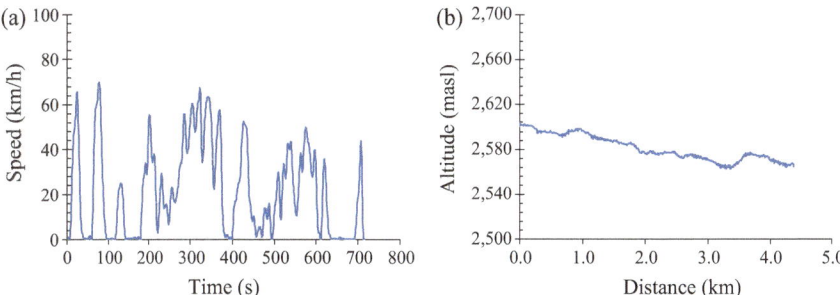

Fig. 10.3 Selected typical driving cycle and its altitude profile

duration, with a maximum speed of 69.62 km/h, an average speed of 21.94 km/h, and a distance of 4.37 km.

10.5   Conclusions

Driving cycles are pivotal in assessing existing vehicles, often conducted on a chassis dynamometer. However, modeling and simulation facilitate a more comprehensive analysis of energy demands by including typical driving cycles with an altitude profile. This, in turn, enables a detailed examination of energy consumption, aiding in the configuration of powertrains and the evaluation of alternative powertrains to enhance overall performance.

Following the Minimum Weighted Differences for Characterization Parameters (MWD-CP) methodology, selecting typical driving cycles from actual driving measurements in a city or region becomes feasible. Defining the boundaries for valid trips is crucial, and in this work, specific parameters are proposed, specifically the cumulative energies associated with each load in the vehicle's driving direction. These parameters directly impact energy demand, making them pivotal in the energy demand analysis.

As a forward-looking consideration, it is crucial to broaden the scope of the study by evaluating a more significant number of vehicle samples in the future. This approach aims to validate and refine typical driving cycles tailored for specific cities or regions. The continuous exploration of diverse vehicle samples ensures the robustness and applicability of the proposed methodology in driving cycle selection and energy demand analysis.

Author Contributions Daniel Cordero-Moreno: Conceptualization, Methodology, Supervision, Investigation, Formal analysis, Writing—review and editing, funding acquisition. **Efrén Fernández**: Writing, Review, Investigate. **Danilo Dávalos**: Data acquisition, Data curation, Formal analysis. **Freddy Vazquez**: Data acquisition, Data curation, Formal analysis.

References

1. GARMIN GPSMAP® 60Cx and 60CSx. http://www8.garmin.com/specs/GPSMAP60CSX0206. pdf
2. Cordero-Moreno, D., Davalos, D., Coello, M., Rockwood, R.: Proposed criteria to determine typical vehicular driving cycles using minimum weighted differences. In: Urban Transport XXIII. WIT Transactions on the Built Environment, vol. 176, pp. 329–337 (2017)
3. Huertas, J.I., Díaz, J., Cordero, D., Cedillo, K.: A new methodology to determine typical driving cycles for the design of vehicles power trains. Int. J. Interact. Des. Manuf. (IJIDeM) **12**, 319–326 (2018). https://doi.org/10.1007/s12008-017-0379-y
4. Cordero-Moreno, D., Davalos, D., Coello, M., Rockwood, R.: Proposed criteria to determine typical vehicular driving cycles using minimum weighted differences. In: Proceedings of the WIT Transactions on the Built Environment, vol. 176, pp. 329–338 (2018)

Case 4: Bucaramanga, Colombia

11

Juan Danilo Molina Martinez, Jessica Gissella Maradey Lázaro,
José I. Huertas, and Victor Romero Cano

Abstract

In recent years, the development of driving cycles has seen a notable increase, driven by the necessity to describe driving patterns, classify driving styles, evaluate fuel/energy consumption, and reduce emissions, particularly in Latin America. Driving cycles represent the instantaneous decisions drivers make to navigate a physical driving environment, known as operation modes. In the case of Bucaramanga city, driving data from sixteen light vehicles and drivers were collected over a five-month period in 2021 using OBD-II technology via Bluetooth connection. This data was gathered without adhering to a specific route, allowing for a comprehensive driving behavior assessment. Utilizing the Micro Trips Fuel Based (MTFBM) method and analyzing nineteen characteristic parameters (CP), a Bucaramanga driving cycle was constructed,

J. D. Molina Martinez · J. G. Maradey Lázaro (✉)
Mechatronics Engineering Program, Faculty of Engineering, Universidad Autónoma de Bucaramanga, Bucaramanga, Santander, Colombia
e-mail: jmaradey@unab.edu.co

J. D. Molina Martinez
e-mail: Jmolina54@unab.edu.co

J. I. Huertas
Sustainable Energy Research Group, Tecnologico de Monterrey, Monterrey, Nuevo León, Mexico 64849
e-mail: jhuertas@tec.mx

V. Romero Cano
School of Computer Science and Informatics, Cardiff University, Cardiff CF24 4AX, United Kingdom
e-mail: romerocanov@cardiff.ac.uk

demonstrating relative differences of the CP values of less than 15%. The results indicate that drivers in the region tend to operate at low speeds and accelerations due to high traffic density.

11.1   Introduction

The automobile fleet in the Metropolitan area of Bucaramanga has experienced a remarkable 87% increase between 2011 and 2018. This figure starkly contrasts with the growth of the city's mobility infrastructure. According to statistics provided by transit and transportation entities, in 2015, the region had 595,373 registered vehicles, which surged to 760,746 by April, 2021, representing a growth of approximately 28%. Consequently, the development of driving cycles emerges as a strategic approach to characterize the influence of driving conditions and modes, represent driving patterns, and minimize fuel consumption.

Most driving cycles developed worldwide serve legislative purposes. In contrast, others address non-legislative needs, considering driving patterns within specific regions or applications, often on defined routes or dynamic driving environments. Additionally, the difference between emissions and fuel consumption observed between driving laboratory tests and on-road tests ranges from 8 to 60% [1].

This case aims to tackle mobility challenges in the metropolitan area of Bucaramanga by addressing the question: What is the typical driving cycle of individuals in the region? Through the implementation of an on-board monitoring system (OBD-II) and a data recording campaign spanning four months, capturing second-by-second data from a sample of 16 vehicles traversing various routes and times of the day, a local driving cycle was developed using the MTFBM method. This approach achieved precision in comparing characteristic parameters of less than 15%, which evaluates the validity of the obtained cycle.

11.2   Case Description

The monitoring campaign was carried out in Bucaramanga during the second semester of 2021. A total of 16 light-duty vehicles were monitored, as shown in Table 11.1, where the relevant technical data of each one is detailed. Furthermore, Figure 11.1 represents the different manufacturers of monitored vehicles, where Chevrolet is the most monitored brand (44%), followed by Renault (25%); finally, the remaining 31% was distributed equally among the additional five brands.

The drivers had the autonomy to travel on any route within the Bucaramanga metropolitan area (Including Giron, Floridablanca, and Piedecuesta municipalities) and at any desired time. The age range with the most participation in the study is between 27

Table 11.1 Technical characteristics of the vehicles used in this study [2]

Manufacturer	Model	Model year	Engine displacement (L)	Gross vehicle weight (kg)	Type of vehicle	Fuel type	Frontal area (m^2)	Drag coefficient (cd)	Rolling coefficient (fr)
Chevrolet	Spark life	2017	1	1,230	Hatchback	Gasoline	1.92	0.34	0.014
Chevrolet	Tracker	2017	1.8	1,806	Pickup truck	Gasoline	2.5	0.352	0.013
Hyundai	Tucson	2016	2	2,030	Pickup truck	Gasoline	2.6	0.34	0.013
Chevrolet	Aveo Sedan	2014	1.4	1,122	Hatchback	Gasoline	2.16	0.36	0.014
Chevrolet	Sail Sport	2015	1.4	1,444	Hatchback	Gasoline	1.76	0.54	0.015
Renault	Sandero	2013	1.6	1,515	Hatchback	Gasoline	2.02	0.35	0.014
Renault	Logan	2014	1.6	1,521	Hatchback	Gasoline	2.65	0.34	0.013
Chevrolet	Optra	2007	1.8	1,720	Hatchback	Gasoline	1.939	0.5	0.017
Chevrolet	Spark GT	2013	1.2	1,368	Hatchback	Gasoline	2.2	0.34	0.014
Renault	Logan	2010	1.6	1,521	Hatchback	Gasoline	2.65	0.34	0.013
Honda	Fit ex	2007	1.5	1,134	Hatchback	Gasoline	1.77	0.32	0.014
Daihatsu	Terios	2016	1.5	1,720	Pickup truck	Gasoline	2	0.34	0.014
Chevrolet	Aveo Sedan	2012	1.4	1,122	Hatchback	Gasoline	2.16	0.36	0.014
Renault	Sandero	2012	1.6	1,515	Hatchback	Gasoline	2.02	0.35	0.014
Volkswagen	Jetta	2019	1.4	1,393	Hatchback	Gasoline	2.5	0.27	0.013
Skoda	Fabia	2018	1	1,035	Hatchback	Gasoline	2.54	0.32	0.014

Fig. 11.1 The brand of the vehicles monitored in this study [2]

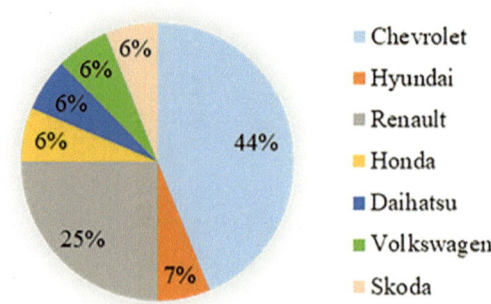

and 59 years old, belonging to the adult group, as seen in Fig. 11.2a. At the same time, regarding gender, it is highlighted that the difference between male and female drivers is 38%, as described in Fig. 11.2b. Finally, driving expertise is balanced in terms of novice drivers with driving experience from 0 to 5 years and regular drivers carrying out this activity from 6 to 10 years, as reported in Fig. 11.2c.

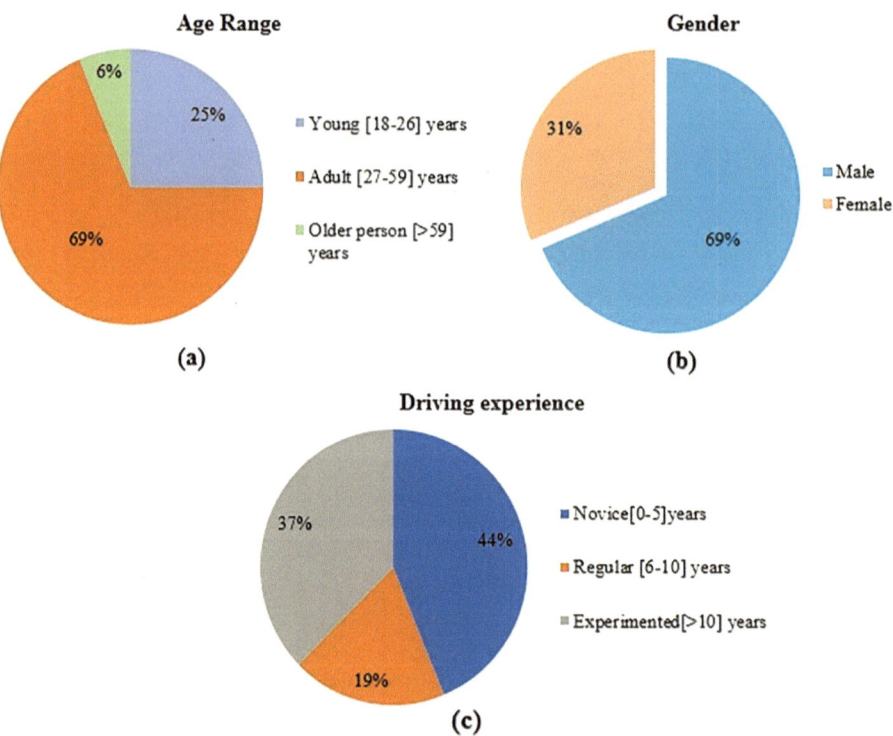

Fig. 11.2 Profile of the monitored drivers in this study [2]

Table 11.2 Characteristics of the monitored region.

City	Area (Km^2)	Altitude (masl)	Distance to Bucaramanga city (km)	Air density (kg/m^3)	Ambient pressure (kPa)	Ambient temperature (°C)
Bucaramanga	165	959	0	1.17	101.3	27
Floridablanca	100.35	926	2			
Girón	475.14	919	7			
Piedecuesta	344	1005	17			

Retrieved from the national administrative department of statistics (DANE)

The population in the Metropolitan Area of Bucaramanga is approximately 1.3M inhabitants in 2024; it has an area of 1,479 km^2 and is located 959 mts above sea level, as shown in Table 11.2.

11.3 Data Gathering and Processing

Figure 11.3 shows the monitoring system, composed of 4 substages: the data acquisition system, the type of signal for sending information, the data logger, and the storage server.

Some characteristics that should be considered when selecting the device to collect data are: the devices must allow the recording and sending of data to an application in real-time, must be able to read the sensors selected by the authors, must have compatibility with most of the vehicles in Colombia, it should not cause any discomfort to drivers when installed due to their size and finally, connectivity must be as stable as possible to avoid loss of information. The OBD-II ELM 327 was selected because it is an alternative with good performance, especially in its Bluetooth version. The variables read through the OBD-II are shown in Table 11.3.

The monitoring campaign was carried out from March to June 2021; 539 trips and 16 vehicles were monitored, traveling approximately 7,000 km and using 800 liters of gasoline.

11.4 Results

As a result of the 539 trips registered for the database of 16 vehicles, a total of 9,897 micro trips were obtained. The clusters are built using the average speed and Euclidean distance as a criterion. Both driving patterns and driving cycles can be described through characteristic parameters (**CPs**) [1, 3], that are metrics based on speed and time, such as average speed and acceleration. In this study, 18 characteristic parameters were defined that

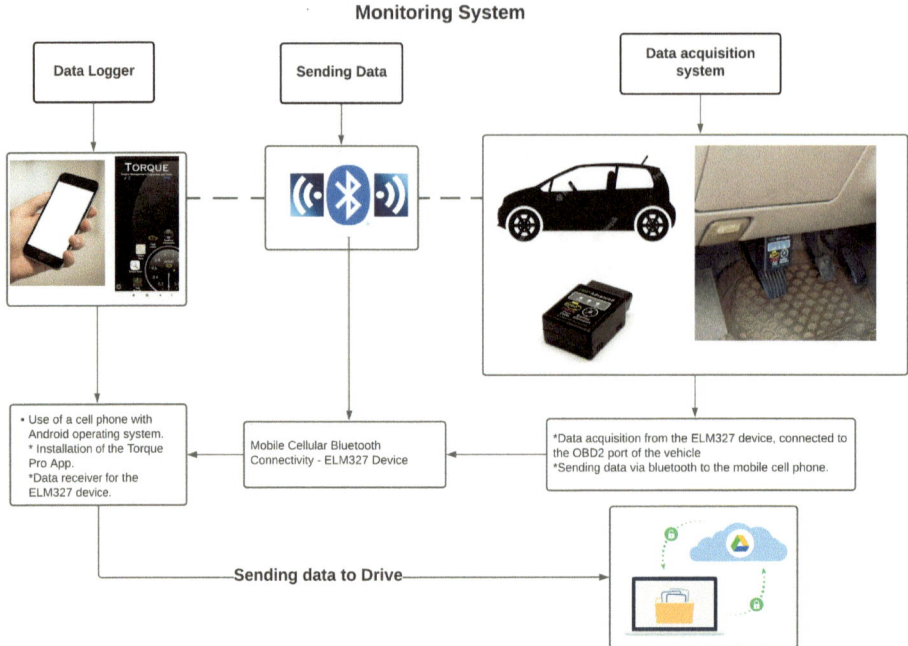

Fig. 11.3 Scheme of monitoring campaign and instrumentation [2]

Table 11.3 Variables recorded by the onboard monitoring devices (OBD-II) in this study [2]

Variable	Unit
Engine load	%
Fuel flow/hour	l/h
GPS latitude y GPS longitude	lat-lon
Liter per 100 km (instant)	l/100 km
Throttle position (manifold)	%
Relative throttle position	%
Manifold pressure	psi
Engine RPM	rpm
Engine coolant temperature	°C
OBD speed	Km/h

describe variables of speed, acceleration, operating modes, dynamics, and fuel consumption. Also, to develop or construct the typical driving cycle for this case, the micro-trips fuel-based method (MTFBM) is selected. This method combines micro-trips (MT) and the deterministic Fuel Based Method (FBM). The base iteration algorithm of this combination is micro trips, unlike its main characteristic parameter to establish that the driving cycle is representative is the specific fuel consumption (SFC); therefore, the condition that establishes that the relative difference (ARD) of the mentioned parameter must be less than 5% concerning that of the data sample to be compared [4]. The candidate driving cycles are validated based on the relative differences between the characteristic parameters of the obtained driving cycle (CPi*) and the characteristic parameters of the driving pattern (CPi). Based on [5, 6], values between [5 and 15%] are acceptable relative differences. Still, this range is also given by the number of selected characteristic parameters and can be adjusted depending on the developed study.

So, the values of the characteristic parameters describing the developed driving cycle are analyzed and presented in Table 11.4. Based on this information, the driving pattern can be determined. For example, the average speed, maximum speed, and percentage of idle provide information about the driving characteristics of the studied region. The driving cycle obtained for the Metropolitan Area of Bucaramanga reflects a low average speed of 5.30 m/s, equivalent to 19 km/h, and a maximum speed of 32.77 m/s, equal to 64 km/h. These values indicate that the city lacks roads that allow speeds greater than 80 km/h and experiences a high demand for road infrastructure due to its fleet of motor vehicles. Furthermore, the percentage of idling reinforces this observation at 23%.

Finally, the DC obtained has a duration of 1,355s, equivalent to 22 min with 35 s. Figure 11.4 shows the typical driving cycle of the Metropolitan Area of Bucaramanga.

11.5   Conclusions

A driving cycle was developed for the metropolitan area of Bucaramanga through second-by-second monitoring, via diagnosis to the board, of a sample of ten vehicles that circulated freely in the city without established schedules or routes.

The micro trip method based on fuel consumption (Micro Trips Fuel Based Method) implemented to construct the driving cycle for the database of 16 vehicles obtained a similarity of the average relative differences of less than 15% of the 16 characteristic parameters that describe the conduction pattern and the 62.5% conduction cycle.

A driving cycle of 1,355 s was obtained, approximately 23 min, where the maximum speed reached was 64 km/h, i.e., a value not exceeding 70 (km/h), primarily due to the urban area monitored region. Another parameter that describes this situation is the idling operation mode (22.9%) [2, 7].

Table 11.4 Results of the characteristic parameters obtained in this study from the implemented methodology [2]

Characteristic parameter (CP's)		Results obtained		
		Driving pattern	Driving cycle	ARD (%)
Maximum speed	Max. speed (m/s)	32.77	17.77	16.22
Average sped	Average speed (m/s)	6.02	5.30	**14.90**
Standard deviation of speed	Std. Dev speed (m/s)	6.03	4.66	16.45
Maximum acceleration	Max + acc (m/s^2)	5.74	2.45	19.49
Maximum deceleration	Max − acc (m/s^2)	−6.48	−3	22.31
Average acceleration	Average + acc (m/s^2)	0.55	0.56	**7.46**
Average deceleration	Average − acc (m/s^2)	−0.59	−0.6	8.31
Standard deviation of acceleration	Std. Dev + acc (m/s^2)	0.42	0.41	10.49
Standard deviation of deceleration	Std. Dev − acc (m/s^2)	0.48	0.46	13.30
Percentage at idle	% Idle	29.15%	22.95%	**2.5**
Percentage accelerating	% Acel	26.80%	29.52%	6.11
Percentage decelerating	% Des	25.28%	28.26%	7.39
Percentage cruising	% Crus	18.77%	19.26%	12.25
Number of accelerations per kilometer	# Accel/Km (km^{-1})	18.16	21.43	19.93
Root mean square	RMS (m/s^2)	0.53	0.55	12.44
Positive kinetic energy	PKE (m/s^2)	0.36	0.38	14.27
Specific fuel consumption	SFC (L/km)	0.11	0.12	**2.43**
Kinetic intensity	KI (km^{-1})	1.04	1.83	36.32

Fig. 11.4 Representative driving cycle for the Metropolitan area of Bucaramanga

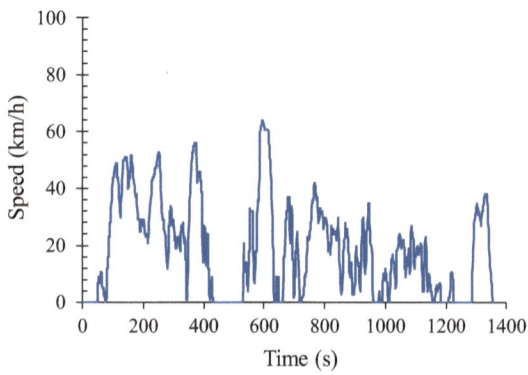

Author Contributions Juan Danilo Molina: Data curation, Conceptualization, Investigation, Formal analysis, Writing—original draft. **Jessica Gissella Maradey Lázaro/José I. Huertas**: Conceptualization, Methodology, Supervision, Investigation, Formal analysis, Writing—review and editing, funding acquisition. **Victor Romero Cano**: Validation, Writing and editing.

References

1. Tong, H.Y., Hung, W.T.: A framework for developing driving cycles with on-road driving data. Transp. Rev. **30**, 589–615 (2010). https://doi.org/10.1080/01441640903286134
2. Molina-Martínez, J.D., Acuña, O.: Clasificación de Estilos de Conducción En El Área Metropolitana de Bucaramanga Con Monitoreo a Bordo (OBD II) En Condiciones Reales de Carretera. https://repository.unab.edu.co/handle/20.500.12749/16806. UNAB: Bucaramanga, Colombia (2022)
3. Tong, H.Y., Tung, H.D., Hung, W.T., Nguyen, H.V.: Development of driving cycles for motorcycles and light-duty vehicles in Vietnam. Atmos. Environ. **45**, 5191–5199 (2011). https://doi.org/10.1016/j.atmosenv.2011.06.023
4. Huertas, J., Giraldo, M., Quirama, L., Díaz, J.: Driving cycles based on fuel consumption. Energies (Basel) **11**, 3064 (2018). https://doi.org/10.3390/en11113064
5. Arun, N.H., Mahesh, S., Ramadurai, G., Shiva Nagendra, S.M.: Development of driving cycles for passenger cars and motorcycles in Chennai, India. Sustain. Cities Soc. **32**, 508–512 (2017). https://doi.org/10.1016/j.scs.2017.05.001
6. Huertas, J., Quirama, L., Giraldo, M., Díaz, J.: Comparison of three methods for constructing real driving cycles. Energies (Basel) **12**, 665 (2019). https://doi.org/10.3390/en12040665
7. Navarro Quintero, R.A., García Jaimes, R.A.: Desarrollo de Un Ciclo de Conducción Bajo Condiciones Reales En El Área Metropolitana de Bucaramanga. https://repository.unab.edu.co/handle/20.500.12749/15102. UNAB: Bucaramanga, Colombia (2021)

Case 5: Mexico City and Toluca, Mexico

12

Michael Giraldo⦿, Luis Felipe Quirama⦿, Juan C. Castillo⦿, and José I. Huertas⦿

Abstract

Driving cycles are employed to assess energy consumption in electric vehicles or the fuel consumption and pollutant and emission factors exhibited by fuel-powered vehicles in a specific region. However, these estimations should be based on representative driving cycles. At times, it is necessary to construct driving cycles that expressly represent the driving characteristics in a city with a specific vehicle typology. In this study, 15 buses were monitored and operated for 8 months by drivers who routinely conduct these routes for the company, avoiding disruptions in regular operations. This work allowed the establishment of characteristic driving parameters for this type of vehicle in the study regions. Finally, through the Energy Micro-Trips method, it was possible to establish the driving cycles with characteristic parameters of 95% similarity to the trips monitored for these buses in Mexico City and Toluca.

M. Giraldo (✉) · J. C. Castillo
Industry, Materials and Energy Area, School of Applied Sciences and Engineering, Universidad EAFIT, Medellin, Antioquia, Colombia
e-mail: mgiral36@eafit.edu.co

J. C. Castillo
e-mail: jccastillh@eafit.edu.co

L. F. Quirama
Sustainable Mobility Unit, United Nations Environment Programme, Ciudad de Panamá, Panamá
e-mail: luis.felipe@un.org

J. I. Huertas
Sustainable Energy Research Group, Tecnologico de Monterrey, Monterrey, Nuevo León, Mexico 64849
e-mail: jhuertas@tec.mx

© The Author(s), under exclusive license to Springer Nature Switzerland AG 2025
J. I. Huertas (ed.), *Fundamentals of Driving Patterns and Driving Cycles*, Synthesis Lectures on Mechanical Engineering, https://doi.org/10.1007/978-3-031-76863-7_12

12.1   Introduction

Transport in Mexico is a crucial sector of decarbonizing the country's economy. According to the International Energy Agency, in 2021, transportation consumed about 35% (159,514 TJ) of the total energy consumed in the country and was responsible for 30% of CO_2 emissions (105.3 MT) [1, 2]. In 2022, the INEGI registered nearly 55 million vehicles in circulation throughout the country, of which 5.6 million circulate in Mexico City, constituting about 2 vehicles per household. According to the Mexico City Emissions Inventory produced by the Mexico City Environment Secretariat (SEDEMA), in 2018, 22,480,079.3 tons of CO_2 were emitted, of which about 71% were made by mobile sources.

To achieve the national and international goals and commitments signed by Mexico, it is necessary to reduce energy consumption, greenhouse gas emissions, and short-lived pollutants produced by the road transport sector. To achieve this, the energy efficiency of the vehicle fleet must be improved, and engine technologies that generate fewer emissions must be adopted. Driving cycles play a fundamental role in understanding driving patterns in urban and suburban regions of Mexico City and their impact on energy consumption and vehicle emissions. The local driving cycles are considered a signature of the driving patterns of a city or region. The driving pattern varies from city to city and region [3]. When tested under their representative driving cycle, changes in the driving patterns also generate variations in the vehicles' emissions and energy consumption.

12.2   Case Description

For this case, we select the region located over a flat area west of Mexico City and Toluca. The main roads cover a total distance of 11.5 km and 18.8 km, respectively, notable for their high traffic, and they are in a flat region of 2255 m above sea level and 2611 m above sea level. The road speed limit is 60 km/h, the average road grades are 1.4% and 1.8%, and the max road grades are 5.2% and 9.0%, respectively. Figure 12.1 presents the routes driven by bus operators in Mexico City and Toluca.

On the other hand, we decided to monitor a vehicle fleet with the same emission control technology and similar maintenance conditions to eliminate the effects of their variations from our results. The buses were make-model BUSSCAR Vistabus with Irizar body frames, built in 2012. An extended description of the monitored vehicle technical characteristics on this route is presented in Table 12.1. The frontal area, drag coefficient, and rolling resistance coefficient were calculated following the procedures outlined in the literature [4].

While monitoring the fleet, the vehicle's latitude, longitude, and speed were systematically measured at 1 Hz through the high-precision GPS device. The onboard Diagnostics (OBD) system was employed to procure real-time fuel consumption data from the Engine

(a)

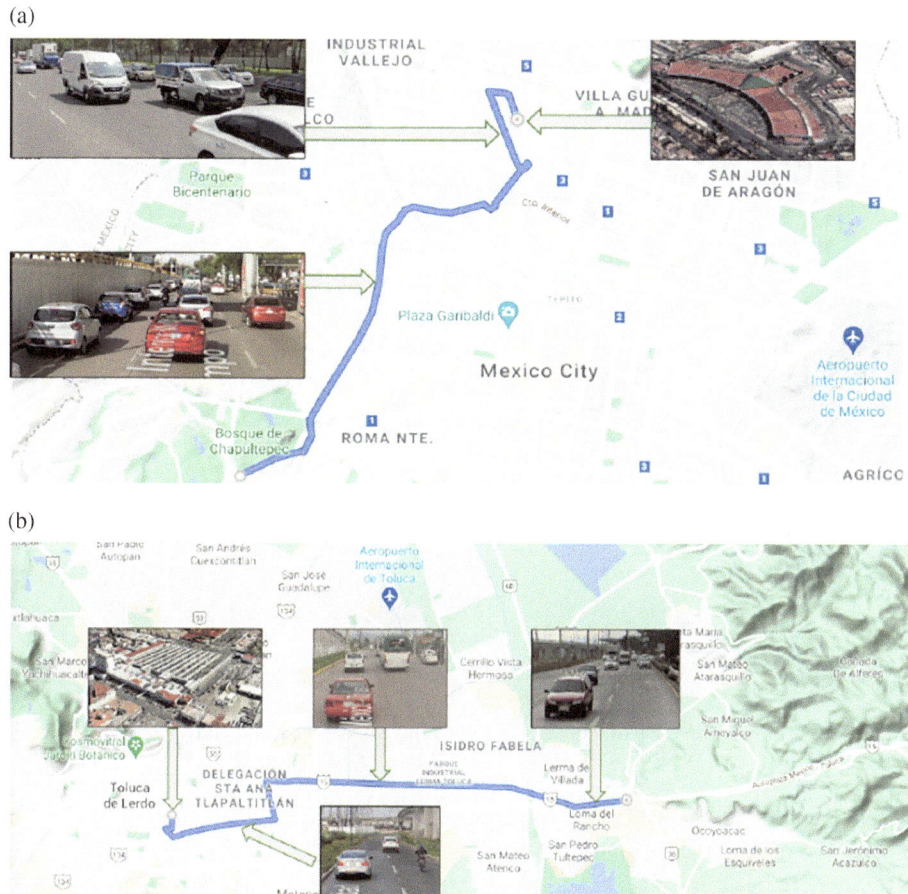

(b)

Fig. 12.1 The route considered in this study is as follows: **a** Mexico City, and **b** Toluca [5]

Control Unit (ECU), which determines instantaneous fuel consumption by monitoring the opening time of the fuel injectors.

For monitoring tailpipe emissions of CO, CO_2, NO, and NO_2 at a rate of 1 Hz, a Portable Emission Measurement System (PEMS) was employed. The PEMS was the SEMTECH ECOSTAR model, with two modules, the SEMTECH-FEM and SEMTECH-NOx. The SEMTECH-FEM module measures fuel economy. It also measures CO and CO_2 emissions using a non-dispersive infrared gas analyzer with a resolution of 10 ppm and a range of 0–8% for CO and a resolution of 0.01% and 0–20% for CO_2. The SEMTECH-NOx module measures NO and NO_2 using a non-dispersive ultraviolet gas analyzer with a range of 0–3,000 ppm and 0–500 ppm, respectively, with a resolution of 0.3 ppm for both gases. To ensure accuracy, the manufacturer's prescribed calibration

Table 12.1 Technical characteristics of the vehicles used in this study [7]

Type of vehicle	Parameter	Unit	Value
	Long	m	12.85
	Wide	m	2.6
	Tall	m	3.6
	Capacity	Passengers	49
	Gross vehicle weight	kg	13.85
	Engine	–	Cummins IMS 425
	Number of cylinders	–	6
	Engine displacement	L	10.8
	Power	HP	425
	Torque	Nm	2102
	Traveled kilometers	km	100,000–200,000
	Control emission technology	–	EURO IV
	EGR	–	Yes
	DOC	–	No
	SCR	–	No
	Frontal area	m^2	8.47
	Drag coefficient	–	0.64
	Rolling resistance coefficient	–	0.006
	Applied load	kg	2100

procedure using NIST traceable calibration gas tanks was executed at the commencement and conclusion of each measurement. We followed recommendations from the U.S. Environmental Protection Agency (EPA) for measuring vehicle emissions concentration [6].

12.3 Data Gathering and Processing

Over eight months, a monitoring initiative was undertaken to track the driving variables (speed and time), altitude, and instantaneous fuel consumption of 15 buses in service, with minimal disruption to their regular operation. Buses were driven by the company's regular drivers, providing a non-stop service (they do not stop to pick up or drop off passengers in the middle of the trip). We included trips performed at different seasons of the year, days of the week, weekends included, and hours of the day. To measure tailpipe emissions, the vehicles were loaded with 2,100 kg of water tanks to simulate the weight and inertia of the passengers. Emissions data were collected over two months following the same routes and under similar conditions.

The data quality was verified in three phases:

(i) Trips with less than 90% of data availability were disregarded. GPS data is frequently lost when the vehicle crosses under a bridge or moves along a tunnel, and PEMS data is lost when the device self-calibrates in the middle of a trip.

(ii) Identified outlier data for each trip. They were data with values outside ranges physically possible—for example, negative values for vehicle speed or values of oxygen concentration higher than 21%. There were also data with values outside typical values, like vehicle acceleration with values greater than 3 m/s^2.

(iii) Synchronization of the data from the vehicle's ECU with the emissions data reported by the PEMS. The non-synchronization of data is originated by the instruments' response time differences. Data synchronization was done manually by dephasing each data set until we obtained the maximum correlation coefficient between variables that, according to physics, should be correlated, such as fuel consumption, engine speed (RPM), and emissions.

Upon evaluating the results from the monitoring campaign and conducting a thorough analysis of data quality, the authors constructed two databases comprising 12 monitored trips within Mexico City and another 12 for Toluca. These trips involved simultaneous measurements of mass pollutant emissions, fuel consumption, position, altitude, and speed. Additionally, 46 supplementary trips were incorporated, featuring identical information but without emission data for each database.

12.4   Results

The driving patterns were obtained to analyze the database trips created for Mexico City and Toluca. It is crucial to reiterate that these buses were in regular service during our monitoring, and we took measures to minimize any disruptions to their regular operations. The average values of driving patterns of Mexico City and Toluca regions were described using the 19 Characteristic Parameters (CPs), which were the most employed in the existing literature [8, 9].

CPs represent metrics derived from speed versus time profiles, encompassing factors like average speed, average positive acceleration, and Positive Kinetic Energy (PKE). Table 12.2 presents the CPs and the values obtained during monitoring vehicles in this study. Additionally, the table includes the 95% confidence level for the range of variation associated with each CP. The values of those CPs were calculated from the speed-time data of the 58 monitored trips.

Once the CPs that characterize the trips in Mexico City and Toluca are obtained, the next step involves creating a driving cycle. The driving cycle was constructed using the EBMT method [10, 11]. In this method, a series of micro-trips (segments of journeys that start and end with a speed of zero) are concatenated based on their average speed and average positive acceleration to form a candidate cycle lasting more than 20 min [12, 13]. The characteristic parameters of the proposed driving cycle must achieve a 95% similarity level compared to the initially calculated characteristic parameters (CPi) for the trips. To

Table 12.2 Characteristic parameters (CPi) were employed to describe driving patterns and values obtained for Mexico City and Toluca

CP	Symbol	Mexico City	Toluca	Units
Maximum speed	Max s	22.3	26.2	m/s
Average speed	Ave s	7.3	10.0	m/s
Standard deviation of speed	SD s	6.9	7.7	m/s
Max acceleration	Max a+	1.3	1.3	m^2/s^2
Max deceleration	Max a−	−2.1	−2.1	m^2/s^2
Average acceleration	Ave a+	0.5	0.4	m^2/s^2
Average deceleration	Ave a−	−0.5	−0.5	m^2/s^2
Standard deviation of acceleration	SD a+	0.2	0.2	m^2/s^2
Standard deviation of deceleration	SD a−	0.4	0.4	m^2/s^2
Idling	% idl	15.1	13.6	%
Acceleration	% a+	32.9	33.8	%
Deceleration	% a−	29.3	29.1	%
Cruising	% cru	22.7	25.9	%
Number of accelerations per km	Accel/km	8.6	6.1	1/km
Root mean square of the acceleration	RMS a	0.5	0.5	m^2/s^2
Positive kinetic energy	PKE	0.4	0.3	m^2/s^2
Speed-acceleration probability distribution	SAPD	N/A	N/A	–
Vehicle specific power	VSP	4.8	7.0	kW/ton
Kinetic intensity	KI	0.8	0.7	1/m
Specific fuel consumption	SFC	0.4	0.4	1/m
Emission index for CO_2	EI CO_2	839.0	749	g/km
Emission index for CO	EI CO	37.2	39.4	g/km
Emission index for NOx	EI NOx	5.0	3.9	g/km

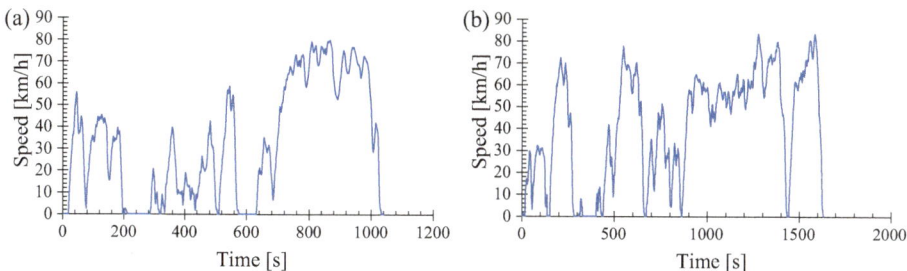

Fig. 12.2 Driving cycles for buses were monitored in this study in **a** Mexico City, and **b** Toluca

achieve this, up to 500 iterations of candidate cycles were conducted, and the one with the smallest difference was selected.

The DC representativeness depends mainly on four factors: (i) the quality and quantity of vehicle operation data, (ii) the DC construction method, (iii) the parameters used to evaluate the DC representativeness, and (iv) the duration of the DC [14]. The obtained driving cycle for the Mexico City and Toluca are presented in Fig. 12.2.

12.5   Conclusions

This work evaluated the driving patterns of the buses that travel through Mexico City and Toluca. This work aims to have the driving cycles that allow a representative evaluation of the energy consumption and exhaust emissions of buses in Mexico City and Toluca.

Through the instrumentation of 15 buses driven for 8 months by company operators and after data analysis, it was possible to evaluate the driving patterns using Characteristic parameters (CPi). In this way, the Driving Cycles for buses in Mexico City and Toluca were created using the Energy Micro-Trips method.

Author Contributions Michael Giraldo Galindo: Data curation, Conceptualization, Investigation, Formal analysis, Writing—original draft. **Luis Felipe Quirama**: Conceptualization, Methodology, Supervision, Investigation, Formal analysis, Writing—review and editing, funding acquisition. **Juan C. Castillo**: Validation, Writing and editing. **José I. Huertas**: Conceptualization, Methodology, Formal analysis, Investigation.

References

1. IEA: CO_2 Emissions in 2022; OECD (2023). ISBN 9789264611542
2. IEA: IEA World Energy Outlook: Paris, France, 2021, p. 2021. France, Paris (2021)
3. Kamble, S.H., Mathew, T.V., Sharma, G.K.: Development of real-world driving cycle: case study of Pune, India. Transp. Res. D Transp. Environ. **14**, 132–140 (2009). https://doi.org/10.1016/j.trd.2008.11.008
4. Huertas, J.I., Coello, G.A.Á.: Accuracy and precision of the drag and rolling resistance coefficients obtained by on road coast down tests. In: Proceedings of the International Conference on Industrial Engineering and Operations Management, vol. 2017, pp. 575–582 (2017)
5. Quirama, L.F., Giraldo, M., Huertas, J.I., Tibaquirá, J.E., Cordero-Moreno, D.: Main characteristic parameters to describe driving patterns and construct driving cycles. Transp. Res. D Transp. Environ. **97** (2021). https://doi.org/10.1016/j.trd.2021.102959
6. USEPA Determination of PEMS Measurement Allowances for Gaseous Emissions Regulated Under the Heavy-Duty Diesel Engine In-Use Testing Program, p. 10024 (2008)
7. Giraldo, M., Huertas, J.I.: Real emissions, driving patterns and fuel consumption of in-use diesel buses operating at high altitude. Transp. Res. D Transp. Environ. **77**, 21–36 (2019). https://doi.org/10.1016/j.trd.2019.10.004
8. Tong, H.Y., Hung, W.T.: A framework for developing driving cycles with on-road driving data. Transp. Rev. **30**, 589–615 (2010). https://doi.org/10.1080/01441640903286134
9. Barlow, T., Latham, S., McCrae, I.S., Boulter, P.: A Reference Book of Driving Cycles for Use in the Measurement of Road Vehicle Emissions. TRL Published Project Report (2009)
10. Lin, J., Niemeier, D.A.: An exploratory analysis comparing a stochastic driving cycle to California's regulatory cycle. Atmos. Environ. **36**, 5759–5770 (2002). https://doi.org/10.1016/S1352-2310(02)00695-7
11. Galgamuwa, U., Perera, L., Bandara, S.: Developing a general methodology for driving cycle construction: comparison of various established driving cycles in the world to propose a general approach. J. Transp. Technol. **05**, 191–203 (2015). https://doi.org/10.4236/jtts.2015.54018
12. Giraldo, M., Quirama, L.F., Huertas, J.I., Tibaquirá, J.E.: The effect of driving cycle duration on its representativeness. World Electr. Veh. J. **12** (2021). https://doi.org/10.3390/wevj12040212
13. Zhang, X., Zhao, D.J., Shen, J.M.: A synthesis of methodologies and practices for developing driving cycles. Energy Procedia **16**, 1868–1873 (2011). https://doi.org/10.1016/j.egypro.2012.01.286
14. Brady, J., O'Mahony, M.: Development of a driving cycle to evaluate the energy economy of electric vehicles in urban areas. Appl. Energy **177**, 165–178 (2016). https://doi.org/10.1016/j.apenergy.2016.05.094

Oscar S. Serrano-Guevara⬤, Juan P. Jiménez, and José I. Huertas⬤

Abstract

Monterrey metropolitan area (MMA) is among the three main cities in Mexico, one of the largest countries in Latin America; additionally, this city has one of the main automotive clusters in the region, so the evaluation of the energy performance of vehicles under real operating conditions in the area is relevant. Two studies determined the typical driving cycle for light vehicles monitored in the region, one to evaluate their real energy efficiency (~26%), with a cycle of 1,227 s, 8.96 km, and 0.078 L/km of specific fuel consumption. In the other, three driving cycles were obtained for the sizing of the electric drivetrain of a passenger bus from a holistic perspective; the typical cycle lasts 1,261 s, 5.25 km, and 0.218 L/km of specific fuel consumption.

13.1 Introduction

Encouraging an efficient use of energy in the transportation sector will have a significant impact on city development.

O. S. Serrano-Guevara · J. P. Jiménez · J. I. Huertas (✉)
Sustainable Energy Research Group, School of Engineering and Science, Tecnologico de Monterrey, Monterrey 64700, México
e-mail: jhuertas@tec.mx

O. S. Serrano-Guevara
e-mail: oserrano@tec.mx

J. P. Jiménez
e-mail: A01281906@tec.mx

© The Author(s), under exclusive license to Springer Nature Switzerland AG 2025   157
J. I. Huertas (ed.), *Fundamentals of Driving Patterns and Driving Cycles*, Synthesis Lectures on Mechanical Engineering, https://doi.org/10.1007/978-3-031-76863-7_13

The energy efficiency of vehicles is determined by fuel economy or specific fuel consumption; onboard data under real operating conditions determine this information through telematics systems that monitor physical variables such as distance, speed, position, and fuel consumption, among others.

As explained in prior sections of this book, monitoring vehicle operation variables for a relatively long period (>1 month) can help determine the typical driving cycle of a specific region and allows the use of this cycle for several applications, i.e., evaluating energy performance, tailpipe emissions, and design of more efficient powertrain configurations.

13.2   Case Description

Monitored vehicle data was used in [1] to determine the real energy efficiency of vehicles operating in several cities of Latin America. Several private vehicles operating under real operating conditions in the Monterrey Metropolitan Area (MMA) were monitored. They determined that, on average, the energy efficiency of these gasoline passengers' cars is between 22 and 25%. Additionally, the driving pattern was also evaluated, and, in this way, the typical driving cycle was determined.

In another work, [2] authors described a holistic methodology based on a multi-objective optimization to define the powertrain configuration (motor, battery, and gearbox) and the energy management strategy for electric vehicles that satisfy the current diesel-based vehicle operation while minimizing TCO and CO_2 emissions, and maximizing energy efficiency, acceleration capacity, and top speed. Aiming to define a tailored power-train for a specific application, authors considered the best (ecofriendly/gentle) and worst (aggressive) scenarios of driving conditions, in addition to the typical (normal) scenario, which is the scenario generally considered in the literature when defining the driving patterns and constructing driving cycles, which from this point on will be referred to as low SFC, high SFC, and typical SFC scenarios.

Figure 13.1 shows the methodology scheme applied in this case; several types of vehicles were monitored through OBD in MMA, located in the northeast of Mexico, a city, known for the significant presence of the automotive industry and its climate, hot semi-arid most of the year. In addition, it is considered among the three main cities in Mexico (along with Guadalajara and Mexico City).

The National Institute of Statistics and Geography (INEGI) reported that in MMA (composed of 15 municipalities), there were 2,488,300 vehicles in 2022, which represents a motorization rate of 0.46 vehicles per capita, and 93% of Nuevo Leon State, so it is well known there are severe traffic conditions during the day.

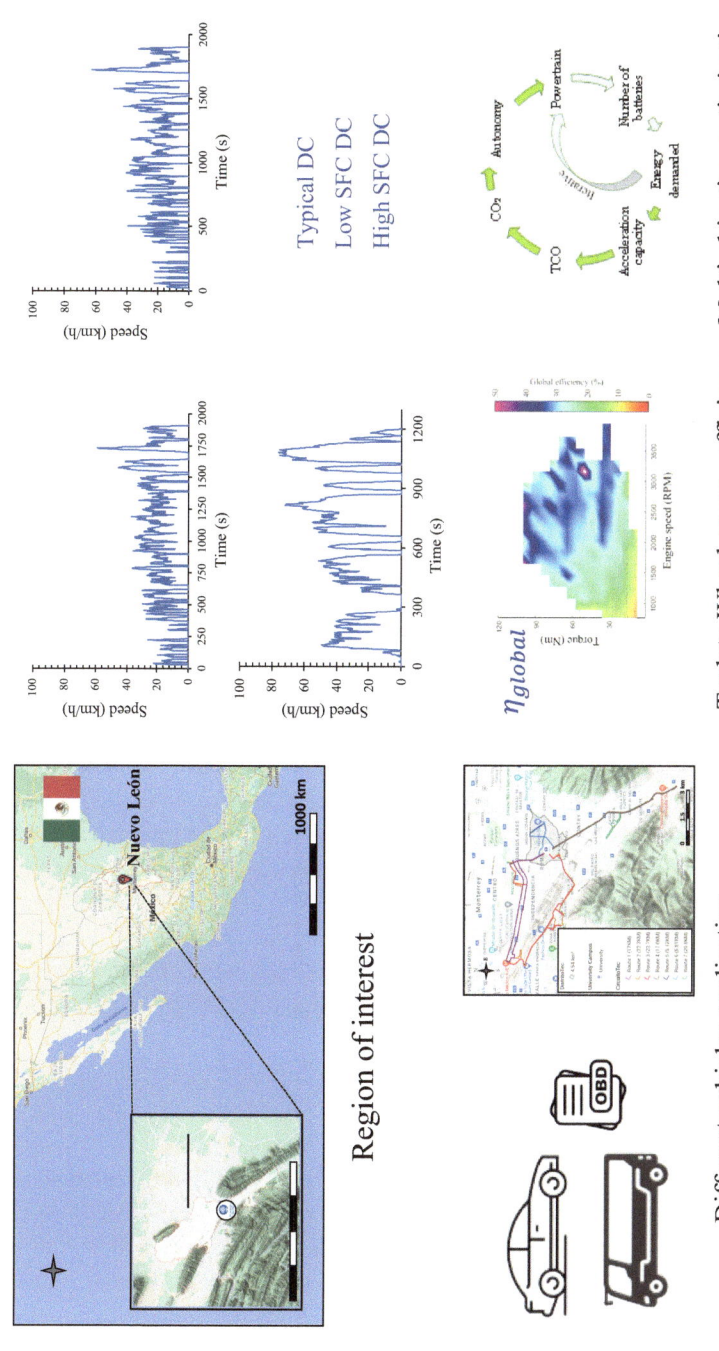

Fig. 13.1 Graphical summary of the methodology applied to the Monterrey driving cycle case

13.3   Data Gathering and Processing

Table 13.1 describes a total of 10 vehicles that were monitored for a 3-month period. This vehicle fleet is composed of passenger cars and passenger vans operating in MMA. At the same time, details of each monitoring campaign are reported in Table 13.2.

For the quality analysis of the collected data, the speed data when the vehicle is off was discarded. This fact is identified when the engine speed operation variable is reported under an idling range (<500 RPM). Furthermore, it was verified that the reported speed values are physically possible and that the reported position is within the geographical area of MMA.

Table 13.2 indicates that approximately 300 k second-by-second data points were collected and used to determine the driving patterns. For the case of the driving cycle, the Energy Based Micro-Trips Method (EBMT) reported in [3] was applied. The different Micro-trips-based methods use absolute relative differences between characteristic parameters (CPs) to get the representative driving cycle; the most influential parameters for selecting the driving cycle are idling percentage, average speed, average positive acceleration, and specific fuel consumption [4]. When the absolute difference is in a 5–10% range with respect to the driving pattern, the generated candidate cycle is selected.

13.4   Results

Figure 13.2 shows the different typical driving cycles determined for MMA. A vehicular traffic condition is observed between congested, due to the high percentage of idling time (>20%), and relatively low circulation speed and fluid due to the regions where the speed is greater than the urban limit of 50 km/h which can be inferred with avenue or highway circulation. Additionally, CPs given in Table 13.3 show that average speed, average positive acceleration, idling percentage time, and specific fuel consumption have the lowest average relative difference (<10%) comparing the obtained driving cycle and the driving pattern considered as the whole time-speed data series. Given this fact, the obtained driving cycles are representative of the MMA and its different applications.

13.5   Conclusions

EBMT method was applied to determine the typical driving cycle for different applications of vehicles operating in the Monterrey Metropolitan Area, get the real energy efficiency of vehicles, and propose and optimize a more efficient powertrain configuration under different scenarios (low, typical, and high specific fuel/energy consumption), including environmental, energy, and economic impacts.

Table 13.1 Technical specifications of monitored vehicles in MMA.

Work	Qty (#)	Manufacturer and model	Model year	Fuel	Reference classification	GVWR (t)	Engine displacement (L)	Engine power (hp @ rpm)	Engine torque (Nm @ rpm)
This work	1	Volkswagen Tiguan	2014	Gasoline	PC–LDV	2.34	2.0	200@1600	280@1700
	1	Kia Rio HB	2017			1.64	1.6	138@6300	167@4800
	1	Volkswagen Jetta	2019			1.97	1.4	150@5000	184@1500
	1	Seat Ateca	2019			1.93	1.4	150@5000	184@3500
	1	Toyota Yaris	2023			1.60	1.5	105@6000	138@4200
	1	Mitsubishi L200	2010			2.85	2.5	131@4000	314@2000
[2]	4	Mercedes-Benz Sprinter	2011	Diesel	PV–LDV	3.50	3.0	188@3800	305@1200

Table 13.2 Details of monitoring campaign of vehicles in Monterrey

Detail	Value	Values
Work	This work	[2]
Trips	147	31
Monitoring period	3 months	3 months
Since, to	April–June 2023	2017–2018
No. of vehicles (#)	6	4
Collected data (#)	231,500	68,310
Distance (km)	2,465	357.38
Used fuel (L)	187.6	77.97
Specific fuel consumption (L/km)	0.076	0.218

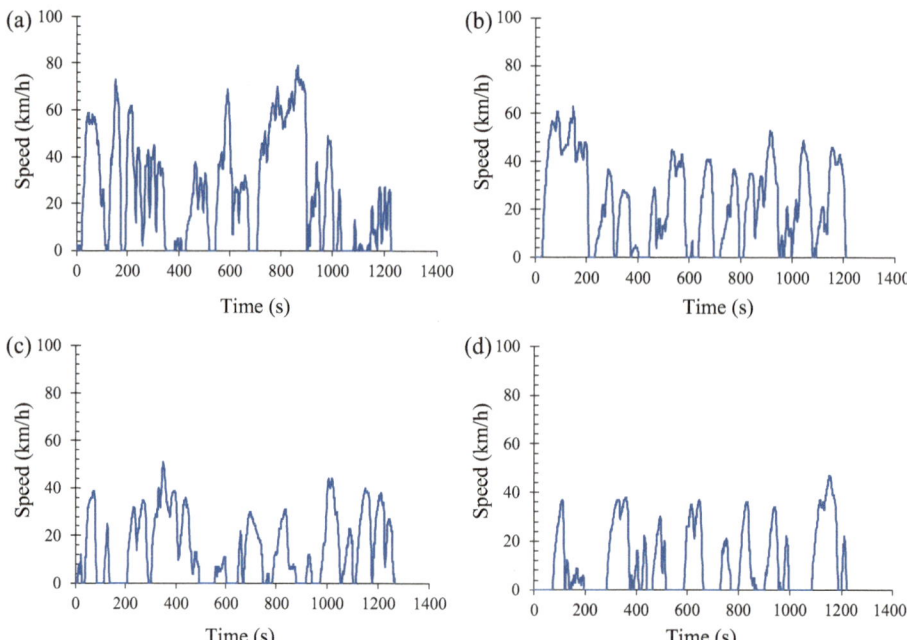

Fig. 13.2 Driving cycles for Monterrey metropolitan area: **a** light-duty vehicles passenger cars reported in Serrano-Guevara et al. [1], **b** low SFC driving cycle for LDV passenger van, **c** typical SFC driving cycle for LDV passenger van, **d** high SFC driving cycle for LDV passenger van. **b, c,** and **d** driving cycles are reported in Huertas et al. [2]

Table 13.3 Characteristic parameters and average relative differences for MMA driving cycles and driving pattern.

Work	This work			[2]				
Characteristic parameter	Driving cycle	Driving pattern	Relative difference	Low SFC DC	High SFC DC	Typical DC	Driving pattern	Relative differences
Time (s)	1,227	231.5 k	NR	1,210	1,233	1,261	68.31 k	NR
Max speed (km/h)	79.00	114	31%	64.01	47.99	51.98	93.30	31%–49%–44%
Average speed (km/h)	26.29	27.12	3%	23.72	14.76	11.16	18.82	26%–22%–41%
Sd of speed (km/h)	22.91	33.41	31%	17.79	13.81	14.05	17.37	2%–20%–19%
Maximum acceleration (m/s^2)	1.57	1.41	11%	1.41	1.48	1.3	1.76	20%–16%–26%
Minimum acceleration (m/s^2)	−2.08	−1.81	15%	−3.89	−2.62	−2.15	−1.73	125%–51%–24%
Average positive acceleration (m/s^2)	0.51	0.47	8%	0.43	0.42	0.43	0.52	17%–19%–17%
Average negative acceleration (m/s^2)	−0.66	−0.25	164%	−0.59	−0.58	−0.59	−0.59	0%–2%–0%
Sd of positive acceleration (m/s^2)	0.40	0.40	2%	0.24	0.23	0.22	0.51	53%–57%–54%
Idling percentage (%)	21.27	20.80	2%	17.4	28.5	44.51	28.55	39%–0%–56%
Acceleration percentage (%)	37.82	34.71	9%	24.06	25.8	16.94	31.40	23%–18%–46%
Deceleration percentage (%)	33.09	31.20	6%	31.77	21.23	23.23	27.90	14%–24%–17%

(continued)

Table 13.3 (continued)

Work	This work			[2]				
Characteristic parameter	Driving cycle	Driving pattern	Relative difference	Low SFC DC	High SFC DC	Typical DC	Driving pattern	Relative differences
Cruising time (%)	7.82	13.30	41%	26.76	23.57	16.22	12.15	120%–94% –33%
SFC (L/km)	0.078	0.076	3%	0.187	0.227	0.300	0.218	14%–4%–38%

Author Contributions Oscar Serrano-Guevara: Data curation, Conceptualization, Methodology, Investigation, Formal analysis, Writing—original draft. **Juan P. Jiménez**: Data curation, Methodology, Investigation, Formal analysis, Writing—original draft. **José I. Huertas**: Conceptualization, Methodology, Supervision, Investigation, Formal analysis, Writing—review and editing, funding acquisition.

References

1. Serrano-Guevara, O., Huertas, J.I., Giraldo, M.: Real Energy Efficiency of Vehicles. Manuscript in Preparation (2024)
2. Huertas, J.I., Mogro, A.E., Jiménez, J.P.: Configuration of electric vehicles for specific applications from a holistic perspective. World Electr. Veh. J. **13**, 29 (2022). https://doi.org/10.3390/wev j13020029
3. Quirama, L.F., Giraldo, M., Huertas, J.I., Jaller, M.: Driving cycles that reproduce driving patterns, energy consumptions and tailpipe emissions. Transp. Res. D Transp. Environ. **82** (2020). https://doi.org/10.1016/j.trd.2020.102294
4. Quirama, L.F., Giraldo, M., Huertas, J.I., Tibaquirá, J.E., Cordero-Moreno, D.: Main characteristic parameters to describe driving patterns and construct driving cycles. Transp. Res. D Transp. Environ. **97**, 102959 (2021). https://doi.org/10.1016/j.trd.2021.102959

Case 7: Saltillo, Mexico

Oscar S. Serrano-Guevara◉ and José I. Huertas◉

Abstract

The certification driving cycles do not necessarily represent the driving conditions of cities in Mexico and Latin America. For this reason, it is important to determine a driving cycle for each case study with a methodology that includes the monitoring of different types of vehicles, with the maximum possible amount of data and available vehicles. Under this premise, the representative driving cycle was determined for the city of Saltillo, an intermediate city in Latin America, characterized by its high industrial development, which generates continuous movements of its inhabitants between the different areas of the city. This driving cycle was determined to highlight the vehicle driving pattern among the fundamental pillars of urban mobility in cities and its intrinsic relationship with energy consumption and vehicle emissions. A speed-time series of nearly 1,198 s, 26% idle time and 0.27 L/km of specific fuel consumption was determined as a representative driving cycle with the energy-based microtrip method.

O. S. Serrano-Guevara · J. I. Huertas (✉)
Sustainable Energy Research Group, School of Engineering and Science, Tecnologico de Monterrey, Monterrey 64849, México
e-mail: jhuertas@tec.mx

O. S. Serrano-Guevara
e-mail: oserrano@tec.mx

14.1   Introduction

In Mexico, transportation plays a key role in the efficient use of energy, it is perhaps the activity with the highest energy-saving potential, in the short and long term. An obvious direct correlation exist between pollutants and greenhouse gases emissions, and the overall road transportation energy consumption [1]. Thus, encouraging an efficient use of energy in this sector will have a significant impact on cities development.

Driving cycles (DCs) are used to evaluate the energy performance of vehicles. The best known DCs in Latin America are those used by Environmental Protection Agency (EPA) in United States for the certification of new vehicles. However, several authors mention that these DCs do not necessarily represent real driving conditions of the region of interest. Therefore, it is important to determine typical DCs for the main metropolitan areas of Mexico. However, few works report DCs in Mexican cities, where the development of methodologies and determination of DCs for the Mexico-Toluca region reported in [2–10] is highlighted. In literature, an additional work done by other authors is reported, Martínez-Martínez et al. [11] mention the use of a DC in the metropolitan area of Monterrey for the energy evaluation of heavy-duty vehicles with a specific fuel, however, the speed vs time series of the cycle is neither reported nor plotted.

14.2   Case Description

Huertas et al. [12] proposed a comprehensive methodology to carry out the assessment of mobility systems in Latin American countries (Fig. 14.1), they used the most recent advances in information technologies, including the use of web map services, low-cost emission tests, and telematics data. From this last point, the determination of driving patterns is included, which, as previously mentioned, is expressed in the form of characteristic parameters, speed-acceleration frequency distribution (SAFD), vehicle specific power, and typical driving cycles.

In the aforementioned work, the assessment of driving patterns is stated as a pillar for the evaluation of key performance indicators for the assessment of sustainable mobility in Latin American cities.

For this case study, the city of Saltillo was evaluated. It is located in the northeast of Mexico, with and average altitude of 1,592 meters above sea level, it exhibits a hot semi-arid climate, with well-defined four seasons. Saltillo is mostly a flat region surrounded by mountain chains running from north to south on both sides. Saltillo has an urban area of around 240 km^2, with a population over 864,062 inhabitants (census year 2018), which results in an urban population density of around 3,600 hab/km^2. Furthermore, it is considered one of the most industrialized areas of the country, with one of the largest automotive clusters in Mexico.

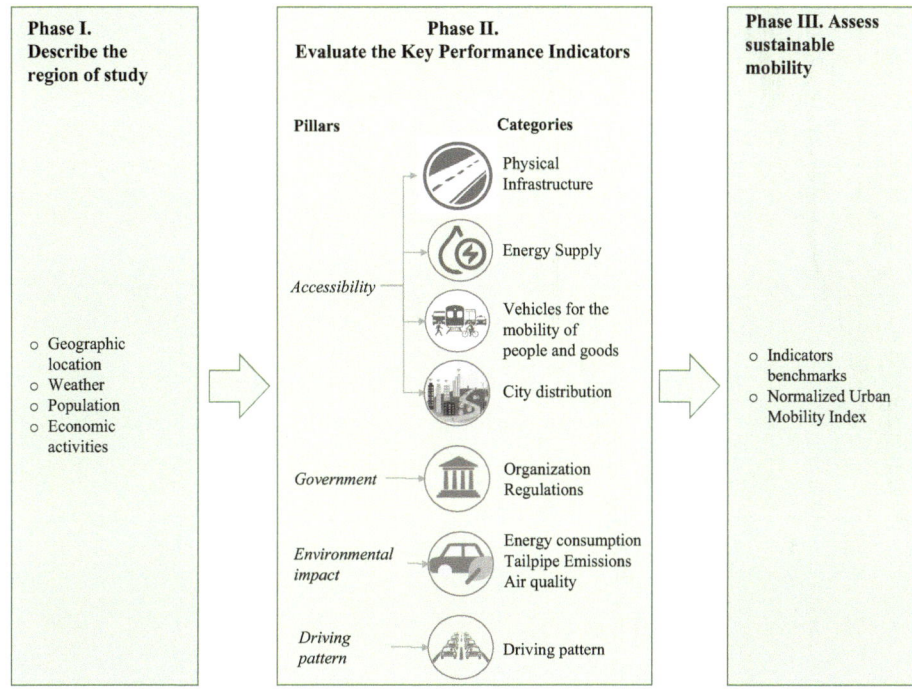

Fig. 14.1 Methodology phases proposed to assess sustainable mobility in Latin American cities (Huertas et al. 2021)

The Secretary of Communications and Transportation (SCT) reported that in Saltillo there were 308,700 vehicles in 2020, which represents a motorization rate of 0.31 vehicles per capita. As reported in section 4.5 of this book, monitoring as much as possible number of vehicles is important to get a typical driving cycle.

14.3   Data Gathering and Processing

A total of 10 vehicles and 583 trips were monitored for a 16-month period. This vehicle fleet is composed by passenger cars, light-duty trucks, and heavy duty-vehicles operating between the urban and the industrial area of Saltillo. These vehicles are described in Table 14.1. This fleet corresponds to different companies' owners that use telematics systems to improve their energy efficiency, safety, and productivity. The service is provided by Metrica Movil Inc. Corp. through Geotab platform. This company uses the GO9 device, which collects the speed-position data with variable frequency. Details of the monitoring campaign are shown in Table 14.2.

Table 14.1 Details of vehicles monitored in Saltillo

No	Manufacturer and model	Model year	Fuel	Reference classification	Gross vehicle weight (t)	Engine displacement (L)	Engine power (hp @ rpm)	Engine torque (Nm @ rpm)
02	International Durastar 4300	2019	Diesel	Class 8–HDT	19	7.6	250@2300	895@1400
03	International Durastar 4400	2018	Diesel	Class 8–HDT	27.5	8.6	300@2000	1290@1200
04	International Durastar 4400	2019	Diesel	Class 8–HDT	27.5	8.6	300@2000	1290@1200
05	International Durastar 4400	2019	Diesel	Class 8–HDT	27.5	8.6	300@2000	1290@1200
06	International Prostar +	2016	Diesel	Class 8–HDT	54	14.9	450@1800	2240@1200
Beat	Chevrolet Beat	2018	Gasoline	PC–LDV	1.5	1.2	82@6400	109@4800
L200	Mitsubishi L200	2014	Diesel	Class 1–LDT	2.5	2.5	135@4000	324@2500
Yaris	Toyota yaris	2018	Gasoline	PC–LDV	1.6	1.5	106@6000	132@4200
07	International Prostar +	2012	Diesel	Class 8–HDT	54	14.9	450@1800	2240@1200
Hiace	Toyota Hiace	2021	Diesel	PV–LDV	3.3	2.8	174@3400	420@1600

Table 14.2 Details of monitoring campaign of vehicles in Saltillo

Detail	Value
Trips	583
Monitoring period	17 months
Since, to	Feb 2019–June 2020
No. of vehicles (#)	10
Collected data (#)	1.04 M
Distance (km)	6.06 k
Used fuel (L)*	212.46
Specific fuel consumption (L/km)	0.092

* Not all trips reported fuel consumption

For the quality analysis of the collected data, the speed data when the vehicle is off was discarded, this is identified from a 0/1 variable reported by telematics system. Furthermore, it was verified that the reported speed values are physically possible and that the reported position is within the geographical area of Saltillo. Subsequently, a linear interpolation was performed to obtain the data with a frequency of 1Hz, as reported in [13], to determine the driving pattern, expressed as: characteristic parameters, speed-acceleration frequency distribution (SAFD), vehicle specific power distribution (VSP), and driving cycle.

As stated in Table 14.2, approximately 1.04 M second-by-second data points were collected and used to determine the driving pattern. For the case of driving cycle, the Energy-Based Micro-Trips Method (EBMT) reported in [3] was applied. The different Micro-trips-based methods use absolute relative differences between characteristic parameters (CPs) to get the representative driving cycle, the most influential parameters for selecting the driving cycle are idling percentage, average speed, average positive acceleration, and specific fuel consumption [4]. When the absolute difference is in a 5–10% range with respect to the driving pattern, the generated candidate cycle is selected.

14.4 Results

Figure 14.2 shows the different ways of expressing the driving patterns for the city of Saltillo. A vehicular traffic condition is observed between congested, due to the high percentage of idling time (>25%), and fluid, due to the regions where the speed is greater than the urban limit of 50 km/h. Additionally, characteristic parameters given in Table 14.3 show that average speed, average positive acceleration, idling percentage time, and specific fuel consumption have the lowest average relative difference (<10%) comparing between the obtained driving cycle and the driving pattern considered as the

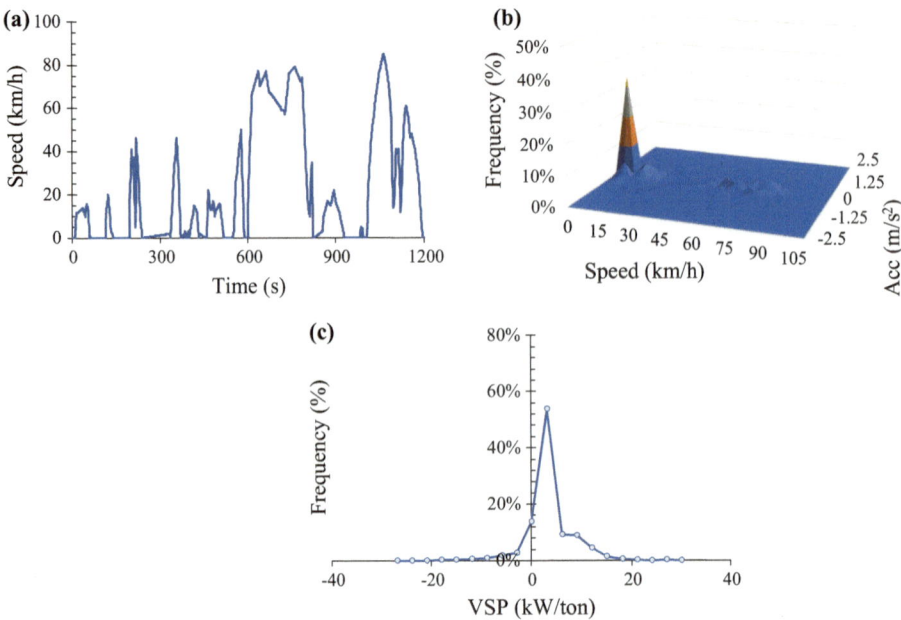

Fig. 14.2 Driving pattern for city of Saltillo expressed as **a** representative driving cycle, **b** speed-acceleration frequency distribution (SAFD), **c** vehicle specific power distribution (VSP)

whole time-speed data series. Given this fact, we conclude that the obtained driving cycle is representative for Saltillo city.

14.5   Conclusions

Driving patterns are important for the assessment of the mobility conditions within any city, thus it has an implicit connection with energy efficiency and vehicle emissions. It is an important pillar when evaluating urban mobility systems, as well as when establishing key performance indicators, and sustainable mobility strategies. For this case study, the EBMT method was used to get the representative driving cycle and compare the different characteristic parameters. Those with the least relative difference regarding the driving pattern (whole data) coincided with those reported in other works as the appropriate ones to determine the driving cycle.

Table 14.3 Characteristic parameters and their comparison

Characteristic parameter	Driving cycle	Driving pattern	Relative difference
Time (s)	1,198	1.04 M	Not relevant
Max speed (km/h)	85.00	125	47%
Average speed (km/h)	22.84	21.12	8%
Sd of speed (km/h)	26.96	23.71	12%
Maximum acceleration (m/s^2)	1.34	1.05	21%
Minimum acceleration (m/s^2)	−1.47	−1.32	10%
Average positive acceleration (m/s^2)	0.28	0.31	9%
Average negative acceleration (m/s^2)	−0.35	−0.55	60%
Sd of positive acceleration (m/s^2)	0.32	0.35	10%
Idling percentage (%)	26.34	26.78	2%
Acceleration percentage (%)	22.88	20.68	10%
Deceleration percentage (%)	21.62	17.62	18%
Cruising time (%)	29.17	34.92	20%
SFC (L/km)	0.272	0.292	7%

Author Contributions Oscar Serrano-Guevara: Data curation, Conceptualization, Investigation, Formal analysis, Writing—original draft. **José I. Huertas**: Conceptualization, Methodology, Supervision, Investigation, Formal analysis, Writing—review and editing, funding acquisition.

References

1. Rafael-Morales, M., Cervantes-De Gortar, J.: Reduced consumption and environment pollution in Mexico by optimal technical driving of heavy motor vehicles. Energy **27**, 1131–1137 (2002). https://doi.org/10.1016/S0360-5442(02)00066-X
2. Huertas, J.I., Mogro, A.E., Jiménez, J.P.: Configuration of electric vehicles for specific applications from a holistic perspective. World Electr. Veh. J. **13** (2022). https://doi.org/10.3390/wevj13020029
3. Quirama, L.F., Giraldo, M., Huertas, J.I., Jaller, M.: Driving cycles that reproduce driving patterns, energy consumptions and tailpipe emissions. Transp. Res. D Transp. Environ. **82** (2020). https://doi.org/10.1016/j.trd.2020.102294
4. Quirama, L.F., Giraldo, M., Huertas, J.I., Tibaquirá, J.E., Cordero-Moreno, D.: Main characteristic parameters to describe driving patterns and construct driving cycles. Transp. Res. D Transp. Environ. **97**, 102959 (2021). https://doi.org/10.1016/j.trd.2021.102959
5. Giraldo, M., Huertas, J.I.: Real emissions, driving patterns and fuel consumption of in-use diesel buses operating at high altitude. Transp. Res. D Transp. Environ. **77**, 21–36 (2019). https://doi.org/10.1016/j.trd.2019.10.004

6. Huertas, J.I., Díaz, J., Cordero, D., Cedillo, K.: A new methodology to determine typical driving cycles for the design of vehicles power trains. Int. J. Interact. Des. Manuf. **12**, 319–326 (2018). https://doi.org/10.1007/s12008-017-0379-y

7. Huertas, J.I., Quirama, L.F., Giraldo, M., Díaz, J.: Comparison of three methods for constructing real driving cycles. Energies (Basel) **12** (2019). https://doi.org/10.3390/en12040665

8. Huertas, J.I., Giraldo, M., Quirama, L.F., Díaz, J.: Driving cycles based on fuel consumption. Energies (Basel) **11**, 1–13 (2018). https://doi.org/10.3390/en11113064

9. Giraldo, M., Quirama, L.F., Huertas, J.I., Tibaquirá, J.E.: The effect of driving cycle duration on its representativeness. World Electr. Veh. J. **12** (2021). https://doi.org/10.3390/wevj12040212

10. Huertas, J.I., Quirama, L.F., Giraldo, M.D., Díaz, J.: Comparison of driving cycles obtained by the micro-trips, Markov-chains and MWD-CP methods. Int. J. Sustain. Energy Plan. Manag. **22**, 1–12 (2019)

11. Martínez-Martínez, S., Rosa-Urbalejo, D. de la, Rua-Mojica, L.F., Hernández-Altamirano, R., Mena-Cervantes, V.Y.: Experimental analysis of real-world emissions using ultra-low carbon intensity biodiesel for a light-duty diesel vehicle in Monterrey metropolitan area. Fuel **317** (2022). https://doi.org/10.1016/j.fuel.2022.123408

12. Huertas, J.I., Stöffler, S., Fernández, T., García, X., Castañeda, R., Serrano-guevara, O., Mogro, A.E., Alvarado, D.A.: Methodology to assess sustainable mobility in Latam cities. Appl. Sci. (Switzerland) **11** (2021). https://doi.org/10.3390/app11209592

13. Huertas, J.I., Serrano-Guevara, O., Díaz-Ramírez, J., Prato, D., Tabares, L.: Real vehicle fuel consumption in logistic corridors. Appl. Energy **314** (2022). https://doi.org/10.1016/j.apenergy.2022.118921